AN ORIGIN OF CONSCIOUSNESS ON EARTH, O-THEORY, AND THE WILL TO EQUILIBRIUM IN THE UNIVERSE

Jason Daniel Bailey

ISBN: 9781701792739

I0510281

AN ORIGIN OF CONSCIOUSNESS ON EARTH, O-THEORY, AND THE WILL TO EQUILIBRIUM IN THE UNIVERSE

By

Jason Daniel Bailey

Approaching the third anniversary, or as I like to call it, the third birthday, of my little brain child, I realized at last it was time for a major revision. On November 15, 2016, I published online what I then called A QUANTUM ORIGIN OF CONSCIOUSNESS ON EARTH AND ITS IMPLICATIONS. That work was exactly 9,623 words long, or approximately 32 printed pages. Over the following three years (the first 18 months in particular), that work grew in size and scope to, as of this writing, 40,596 words, or approximately 130 printed pages. The majority of that growth consisted of random phenomena that seemed to find their explanation in the three theories I proposed in that initial work. As I read and studied and sought refutation and disproof for my ideas, I kept finding support instead, dozens and dozens of phenomena in the macro and micro worlds that appeared to make sense if and only if the ideas I proposed were in fact an accurate description of reality. This was and remains a very big "If." As with the initial vision of the origin of consciousness (elaborated on later), almost all of these ideas hit me involuntarily. I was not ruminating or reflecting on these phenomena. The ideas hit me unexpectedly. They flooded in and bombarded my mind quicker than I could write them down or speak them into my voice notes. The paper doubled in size in a matter of months, tripled by the end of 18 months, and quadrupled before I acknowledged it had become, in modern parlance, a "hot mess" in need of revision and organization. I attempted such a project in August 2018. It helped. But since then more ideas have come, and I have also since discovered the work of Claude Shannon and Information Theory (a field that received considerable attention in the most recent additions to my paper). Also, current research and discoveries in fields which I address in this paper convinced me 1.) that the work needed serious revision and organization, and 2.) that my ideas seem more and more to be correct. A dozen thinkers and searchers on the same path as myself have published studies and research findings, none of which refute or disprove my proposals, and all of which appear to support or confirm my ideas. It has felt as if, for years past, the last three in particular, that I have been one step or more ahead of these seekers after a complete description of Nature, possibly having supplied some "final" answers. Possibly not.

Although I have a certain confidence in my ideas, that I may actually have "nailed it," as several of my readers have claimed, the paper itself has several egregious flaws. The first, and most damning in the eyes of real scientists and science journals, is its format. It is not a science paper. It is not "scholarly," as Huping Hu so kindly pointed out in his rejection of a scaled-down version of "Section I: Consciousness." I make no apologies for such an error, as my background is literature, not science. I am an English professor, a bit of a philosopher, and a student of human nature. Rather than force the paper into a form acceptable to current journals publishing articles and reports in the fields which I here address, I have decided to "double down" on what may be an error, and make this work even less a "science" paper, and more of a witness, a testament of sorts, a piece as revealing of myself as it purports to be revealing of the universe.

So, as I revise and condense what is already here, I will also be even more conversational than I was in the original publication. I tried to make it scholarly. I attempted wholeheartedly to squeeze it into an acceptable mold for *Science* or *Nature* or the *Journal of Consciousness Exploration & Research*. I failed. So I am taking it back and making it wholly my own. It's my baby. Its DNA is my own. So it should look and sound as much like me as any natural child I may have helped bring into the world.

As a result, the first change previous readers, and pearl-clutching science purists, will notice is the overt use of the first-person pronoun, "I." This is now a very personal document. It purports no claim to science. It is just my thoughts and reflections on a few of the timeless questions of humanity. It's not even philosophy. It may be no more than an epic prank. How much funnier will it be, should I turn out to be right?

This will not be a complete revision. Much of what appears in the original paper will make its way here intact. That is a result more of laziness on my part than any grand design of producing a work of literary merit or scientific value. Therefore, much of it will be familiar to previous readers, as I am an avid copy and paster. Thank you Alan Turing and Claude Shannon! And Bill Gates.

My only real nod to "real" science is I will keep intact the "Abstract" that has accompanied by brain child for much of its three-year life. I wrote the abstract many months after initial publication. When the paper was only 32 pages long, it was not too much to tackle, and I was not in the mindset of it being a formal "science" paper at the time. Real science papers have abstracts. So I wrote one. Here it is:

ABSTRACT: This paper is divided into six sections, (actually seven, counting a brief "caveat"), comprising the proposal, elaboration, and genesis of three new hypotheses. In Section I, I propose my first hypothesis, an origin of consciousness on Earth. I begin with a brief narrative (a personal vision which, in and of itself, is potentially a fourth hypothesis), speculating on approximately where, approximately when, and approximately how the ingredients of consciousness (and of life itself) came together. I proceed from there to list and describe ten degrees, or levels, of consciousness (Reactivity, Bonding, Instinct, Essential Awareness, Cognition, Permanence, Intention, Foresight, Self-Awareness, and Global Awareness), as they appear on Earth, expressed in a seamless spectrum. Section II elaborates on a phrase I introduce in Section I—variable electron reactive behavior (VERB)—the most basic and irreducible element of consciousness. It is a new term describing previously-established and observed behavior of electrons in response to stimuli. In Section III, following a line of thought sparked by my musings on consciousness, I propose a second hypothesis, a mathematical equation—$0=1$—as a foundational aspect of reality. (I know, "A 0 can never equal a 1!" Nature, however, proves otherwise.) I round out Section III with nearly a score of quantum and cosmological phenomena for which $0=1$, if true, offers an explanation. In the very brief Section IV, I propose my third hypothesis—that a "Will to Equilibrium" is yet another fundamental force of Nature. Section V, by design, is random. Titled "Other Implications," it is a collection of phenomena in both the macro and micro worlds, including new ideas regarding light, black holes, the Big Bang, and a potential unified field theory that is compatible with classical physics, general relativity, and quantum mechanics, and the final fate of the universe. This hodgepodge of phenomena adds to the list of implications in Section III supporting the equation 0=1, as well as those accounting for the other two hypotheses—the origin of consciousness and the "Will to Equilibrium." More importantly, Section V is where I introduce and refined what I now call O-Theory. Section VI is pure reflection—consisting of autobiography, details of the intellectual journey that led to the creation of this paper, and speculations on future applications of the ideas contained herein.

Reading over it just now, I cannot avoid a smile at the almost inadvertent mention of O-Theory in that next-to-last sentence. Just like the tongue-in-cheek title of the original publication, with its "Implications" hiding what turned out to be the far more important components of the paper, so this passing reference to something called "O-Theory" gives me a chuckle. As if O-Theory were merely a side-thought. And, yes, I call it "theory." Untested, unproven, and virtually unknown, I have every confidence O-Theory will prove true and one day warrant that moniker. Or not.

With the abstract out of the way, I will carry on with another piece already included in the current version of the paper: Section 0: Caveat. I added this piece probably well after a year of initial publication. It is a result of much of the feedback I received from readers in the first ten to twelve months. Mostly attacks more how ludicrous my concepts were than actual criticism or refutation of the actual science I based those concepts on, I realized it was time to make a statement about the work, a statement that went beyond what I had already said in Section VI when I first posted it online. That Section has remained intact almost word for word since I first shared with the world in November 2016. Most likely, it will make its way into this version still virtually unchanged. We shall see.

However, to the caveat, as I call it. I realize a lot of what it contains I am now repeating or paraphrasing in these new opening pages. So be it. I can promise the reader that a lot of changes are coming. A lot of condensing of repeated ideas, culling of areas where my thought has evolved, and excising where I got it flat wrong is coming. But some pieces will remain much the way they first appeared, until such time as a professional editor puts his or her hand to it. I will not be holding my breath for that day. As of this moment, the caveat will remain numbered 0. I retain it for many reasons, not the least of which is that it the first place I drop a few names of seekers whose contributions to my research are invaluable. Moreover, scientists like to see the names of actual scientists in publications purporting to be "science," especially their own names. The paper made it eventually to six sections. That number will most likely increase, as concepts like O-Theory and the parallels to Shannon's Information Theory will now warrant sections all to themselves. For now, here is Section 0:

0. CAVEAT

Roger Penrose, along with Stuart Hameroff, wrote in 2011: "Orch OR [Orchestrated 'objective reduction'] places the phenomenon of consciousness at a very central place in the physical nature of our universe, whether or not this 'universe' includes aeons other than just our own. It is our belief that, quite apart from detailed aspects of the physical mechanisms that are involved in the production of consciousness in human brains, quantum mechanics is an incomplete theory. Some completion is needed, and the DP [Diosi-Penrose] proposal for an OR scheme underlying quantum theory's R-process would be a definite possibility. If such a scheme as this is indeed respected by Nature, then there is a fundamental additional ingredient to our presently understood laws of nature which plays an important role at the Planck-scale level of space-time structure. The Orch OR proposal takes advantage of this, suggesting that conscious experience itself plays such a role in the operation of the laws of the universe" (257).

I read this paragraph for the first time the week of December 19, 2016, just over a month after first publishing the paper I here offer you. What struck me is that the ideas that follow could very well supply the "completion," the "fundamental additional ingredient" that Professors Penrose and Hameroff were calling for in 2011.

I will let the reader, and time, decide.

One problem I foresee, and have already encountered with early readers of this work, is the demand it makes on the reader to abandon specialization and open his or her mind to potentialities made available from the synthesis of fields with which he is familiar and those to which he may be a stranger. As I say in Section VI, this paper is the product of reading and study in over thirty fields of science and humanities. Readers familiar with cosmology may not have a grasp of oceanography. Those well versed in molecular biology may be limited in psychology and theories of mind. Researchers and theorists in quantum mechanics may have a fuzzy or no grasp of volcanology. This is a paper that demands at least a cursory understanding of all of these fields of study in order to grasp the connections I feel I have made. I merely ask that one read the paper thoroughly and not reject it due to incredulity sparked either by one's lack of experience in a particular field or by the biases he or she holds for their particular area of expertise.

Another objection I expect to hear (and have already) is the paucity and virtual lack of research, data, and analysis. In initial drafts, there was none. I have remedied that somewhat in the ensuing year since publication. As I state more at length in my concluding section, this is a work of synthesis. All the research and data has been compiled by intellectual explorers long before I began compiling my own feverish notes. But the lack of original research should not deter the reader from giving my ideas a fair shot. Charles Lyell's *Principles of Geology*, I remind the reader, was also a work of synthesis in which he interpreted the composite meaning of the work and workers that preceded him. I have always had a talent for synthesizing fragments from disparate fields of study into a whole that is greater than the sum of its parts. I believe that is what I have done with this current paper.

This is not a science paper. It is a paper that contains a considerable amount of science. It is not a work of philosophy, though it waxes philosophical at times. It is not a metaphysical or religious paper, yet it raises questions usually relegated to those realms of thought and speculation. It has less in common with Darwin and Einstein and more in common with Democritus or Lucretius. And yet there is not a word of it that is not grounded in established, experimental, and verifiable science, except where it leaps into the purely theoretical. It's a strange paper. I'm the first to admit this. There has been nothing like it published to date. I take pride in its quirkiness. It is, in the final analysis, a very human paper. It is flawed. It is at times meandering. It is simultaneously self-reflexive and stubborn. It is written as much in certainty of the established laws upon which it rests as it is in hope that the bold proposals it offers may someday become established truth. And, as the human who wrote it, I simply hope the reader will give it a chance.

Hold on tight, Dorothy. We are leaving Kansas for good.

I have to admit, those last two sentences still make me giggle. Readers who encountered those words of warning when I first added them to the paper have fallen into two camps: Those who got the joke, smiled, and kept reading, and those who stopped and read not one more word. One reader messaged me that it was good sign my paper started with the prestigious name "Roger Penrose." That reader was also one who did not make it past the Wizard of Oz allusion. So be it.

Approaching Section I: Consciousness, I notice even early drafts of the paper were a bit snarky and unconventional. After the opening sentence of that sections comes, "And here you pause. No need to read further, right?" This conversational tone, this addressing of the reader, this breaking of the "fourth wall" approach is what I see this revision becoming more and more. Though there will be considerable revision, believe me, this is as much a reflection as it is anything. It is me thinking out loud about my own thoughts. A modern-day Montaigne with far less talent.

I will not be taking the time or spending the words to point out areas of actual revision. Previous readers will recognize most places where something has been revised, re-written, condensed, or "disappeared" altogether. Just know, dear new reader, from this point forward, there will be many changes.

I. CONSCIOUSNESS

Consciousness precedes life.

And here you pause. No need to read further, right? You've heard all you need to hear to know this paper is ludicrous and its author clearly out of his mind. Understandably, many a reader will be confounded by this contention–consciousness precedes life–as it runs contrary to every idea of consciousness they have held since they first pondered it. But hear me out. Consider one instance: Debate has waged for decades: Is a virus alive or not? Some say yes; some say no. Some say yes, but only once the virus is nestled within its host. (Oppenheim and Putnam 24). In regards to consciousness, this question is irrelevant. Alive or not, a virus is conscious.

But what is consciousness? The problem with previous treatments, like those by Daniel Chalmers, Danah Zohar, and Marc Seifer (who comes REALLY close to discovering what I here propose), is that at some point, all of them leave the realm of physical science and venture into metaphysics. Chalmers quite proudly rejects what he sees as the reductionist approach. My approach is pure reductionism. Then there are the exceptional writings of Daniel Dennett, which, while avoiding the pitfalls of metaphysics, manage to dazzle with the processes and functioning of consciousness, without ever telling us where it comes from or what it is at its most fundamental level. After 400 or more pages of more than one study, Dennett still seems to say, "But no one really knows what consciousness is. It's a mystery!" Meanwhile, as this paper will show, what many persons refer to as "consciousness" is really self-awareness, a degree of consciousness requiring complex structures evolved over a few billion years. And they are not wrong to call it thus. My question, however, is what is its origin? Where did it come from? What was consciousness before it was "consciousness"? Moreover, at what point do we delineate consciousness from what preceded it? I have a difficult time drawing lines in the proverbial sand and saying this is consciousness, but that is not. On a continuum of decreasing complexity going backwards in time, how far back do we go to pinpoint when and where consciousness first appeared on Earth, however primitive and unlike self-awareness it may seem? My primary approach has been reductionism. Reductionism is so attractive and compelling because it is easier to go backwards than forwards. Guessing forward, at this point, remains elusive. We can tell from where evolution has brought us, not where it is taking us.

As a materialist and a reductionist, after over two decades of study and thought, I was struck by a vision. Literally awakened from a dream in which I "saw" the answer, or at least what I argue here to be the answer, my dream became my hypothesis. And my hypothesis is this:

Consciousness on Earth originates with what I have newly labeled the variable electron reactive behavior (VERB) in atoms prevalent at deep ocean volcanoes and hydrothermal vents on the early Earth (forming carbonic molecules that are the building blocks of crystalline nucleobases and thus for life itself), and increases in complexity and depth of experience as the complexity of the structures and entities exhibiting consciousness also increase over time via natural selection and (recently) cultural drivers.

There are many who would say the preceding paragraph does not meet the definition of a hypothesis, as it makes no testable predictions. This is a "problem" with all three ideas I propose over the course of this paper. I have no doubt that if my ideas are correct that testable predictions can come from them. However, most of what follows is what I call prediction in hindsight. Phenomena we already know to occur, yet have few or no explanations for, are hereafter explained by what I am already brashly calling "theories". By posing several "what if" scenarios, I feel the reader and the science community at large will see that my ideas are not only plausible, they are probable explanations for much of what in this universe currently remains a mystery. In addition, those who would say a dream is no way to go about doing science need only recall the story of Mendeleyev. It was in a dream that he "saw" the elements fall into place into what he would soon name The Periodic Table of Elements (Watson 85). Mendeleyev's dream came after three days of intense study; mine came after thirty years.

My concept of VERB (which I elaborate on in the following section) is based soundly in established science. Empirical experiments prove out that the behavior I ascribe to electrons in indeed fact. I have made nothing up. I merely say that the known behavior of electrons is the basis for what we call consciousness.

Famed Danish physicist Niels Bohr revealed how atoms are formed by the "binding" of electrons and how this "binding" determines the chemical properties of elements formed from those atoms. Electrons generally behave stably unless acted upon. Bohr stated how certain conditions enable electrons to make quantum "jumps" to new orbits, changing the characteristics of the elements. (Watson xviii; NOTE: Unless otherwise specified, all citations from Watson refer to his 2016 work, CONVERGENCE, an invaluable resource in the revision of this paper). My contention is that one such condition causing these possible jumps was the extreme environments of the hydrothermal vents. This will become, hopefully, relevant as we move forward.

Before we proceed, allow me to share a story with you, my own personal vision of what occurred in the nascent days of our planet. As with VERB, what follows is my own speculation based on decades of reading. The groundwork for what follows was laid by other researchers and intellectual explorers. This "narrative" is the conclusion I have drawn from my reading. It is, in brief, how I believe events unfolded in the infancy of this Pale Blue Dot. Anyone taking exception to my version of the story are welcome to propose their own more-plausible alternative.

Four billion years ago, deep in an ancient ocean, at the crease where tectonic plates of two long-vanished continents met, something amazing happened. Magma from under the Earth made its way to the crust and, due to the pressures deep below, found its way into the oceans as lava, volcanic ash, and gases. Among those elements, in abundance, was carbon. Much of this carbon made it into the ocean, where it remains. A similar event was occurring above water, on land, with active volcanoes helping to create the Earth's early atmosphere. Back in the oceans, something very significant for us living today was taking place. The electrons of the carbon atoms were behaving very strangely. Electrons are known to be erratic little buggers. Their state can vary, determining the fate of their particular atom. In the case of the carbon atoms, the extreme volatile conditions that existed at these deep-sea volcanoes and crevices had a quantum electrodynamic (QED) impact, causing the atoms to do one of three things: those carbon atoms that remained too close to the extreme temperatures of the sub-oceanic lava remained erratic components of the magma/lava and did no one much good. Those atoms that found themselves exposed to the dark, frigid waters of the deep ocean became listless and ended whirling about all alone or bonding haplessly with other atoms to little or no significance to later Earth history, or floating up and escaping the oceans to become part of the fledgling atmosphere forming above. However, the third group of carbon atoms found a happy medium, an optimal or "Goldilocks" zone just beyond the volcanic temperatures–not too hot to break down, not too cold to flitter off to anonymity. They struck just the right balance to bond in mass with their fellow atoms, in particular their two closest cousins, nitrogen and oxygen, and their much lighter, but highly abundant, relative, hydrogen. The combination of these four neighbors, of course, form the four nucleotides that form the basis of all life. These molecules were able to settle in the area. Through the combination of spontaneous order and natural selection, these molecules eventually formed the double-helix we know as deoxyribonucleic acid, or DNA. This event exemplifies what Ilya Prigogine calls "dissipative structures" formed by the flux of energy and matter. For more elaboration, see Lane's *The Vital Question* (94-95). For more on spontaneous order, see *Chance and Necessity* by JaquesMonod and *The Origins of Order* by Stuart Kauffman. A new study in Nature Chemistry by Gibard, Bhowmik, and Krishnamurthy suggests that diamidophosphate (DAP) in addition to time and sponaneous order could be the key to DNA's formation, with natural selection determining its success. So minus the DAP, I appear to have some support for this aspect of my theory.

This idea that consciousness and life began at thermal vets near the edges of tectonic plates find support in *The Vital Question* (2015) by Nick Lane, as well as other researchers. Again Watson gives a great summation of Lane's thought: "His view, shared by numerous colleagues, and now thought through to the atomic level and backed up by a number of experiments and observations, is that life on earth began in these hydrothermal vents, which are the areas where energy (radioactive decay) escapes from the center of the planet in the form of heat" (487).

Ilya Prigogine in his 1984 work, *Order Out of Chaos*, argues that "spontaneous organization" occurs in nature and is central to maintaining order. This way Nature appears both highly diverse, and yet highly ordered (71). Stuart Kauffman using mathematical biology, goes to say this same spontaneous order exists in cells and in the ordered morphology of organisms, not merely inorganic molecules. Kauffman gore so far as to use the term "crystallize" in reference to initially disordered systems (173). All of Nature's history has been the emergence, disappearance, and reemergence of entities and species all seeking and finding a form of order. The tendency to spontaneous order is, itself, consciousness. Pitted against other species and the environment in a colossal dance or battle. So,times it is hard to tell the difference. What remains today from the "struggle" are the winners in the game of natural selection. Other entities form and fade. The survivors are those with the most successful organization, best suite to its environment. Spontaneous organization occurs at all levels of nature, yet it is natural selection which gives us what we see around us today.

Watson writes, "evolution proceeded by the development of a series of 'basic plans,' spontaneously organized regularities among the proteins, and this is how the higher-order taxa became established" (443). The predictability (spontaneity) of patterns increases as complexity of the entity decreases back through the lower orders to the molecular and atomic levels.

Prigogine also argues that self-organization of inorganic substances is what eventually led to life. I could not agree more. All he was missing were the thermal vents and the spectrum of consciousness that emerged from them.

In both RNA and DNA, carbon is the primary ingredient. It is for this reason my story focuses on the role of carbon in the origin of consciousness on Earth. It is also why I emphasize, in my title and throughout my paper, the phrase "on Earth." On other planets or celestial bodies, the story could play out differently. Where elements exhibit different properties, due to varying gravity or temperatures, we might discover silicon-based life forms, or even methane-based life forms. However, for now, for our list of characters in the narrative of consciousness on our home planet, our protagonist appears to be carbon.

Clearly this narrative is a gross simplification of what happened. In my rush to publication I failed to elaborate on the billions and billions of chemical processes occurring over hundreds of millions of years, the end products being the results of almost implausibly astronomical numbers of algorithms all occurring at once. Nor do I explore the minutia of QED effects that the extreme temperatures had on constituent particles that would go on to form life itself. For exactly what those effects may comprise, see Gribbin (312). From the dance of atoms swirling near the sub-oceanic vents to RNA and DNA was a process of hundreds of millions of years. From RNA to photosynthesis was hundreds of millions of years with algorithms including—among numberless others—plate tectonics and the emergence of entities exchanging lava heat for solar heat as a source of energy in the resulting shallower waters. To get from chemical compounds and molecules and crystal formations that merely endure to those which actually replicate, from crystal colonies that migrate to photosynthetic slimes to the Ediacara, though taking multiple hundreds of millions of years, is merely a matter of degree, with no distinction along the spectrum from one to another. Erwin Schrodinger himself correctly predicted that genes (later DNA structure) were simply aperiodic crystal formations, as opposed to periodic crystals like salt and ice (Mcafadden and Al-Khalili 205-207). Daniel Dennett, in his latest book, *From Bacteria to Bach and Back*, does an excellent job of filling in my narrative for me with speculation on processes involved, processes I assumed were implied and understood, not in need of explicit delineation. Dr. Dennett lays out these processes while yet not realizing the significance of them to the emergence of consciousness, and therefore, life on Earth. He appears to still be looking, as most researchers are, for that "spark" that was the "beginning" of life. Likewise, Terrence Deacon, in his book *Incomplete Nature: How Mind Emerged from Matter*, coins and describes processes that could very likely "fill the gaps" in my narrative of how we got from VERB to DNA and other complex structures. But also like Dennett, Deacon seems to have missed the significance of his own discoveries by not taking them to their logical conclusion. McFadden and Al-Khalili's engrossing work, *Life on the Edge*, also does a masterful job of filling in the details of life processes, chemical reactions, catalysis, rooted in quantum events. Most explorers in this field seem "stuck" at a point of reduction past which their preconceived notions will not allow them to step. Perhaps the most thorough exploration of the potential paths these processes may have taken can be found in Chapter 3 of Lane's *The Vital Question*. For every entity, or species, we see thriving today, there are untold numbers who did not "make it."

According to Max Perutz and others after him, proteins themselves have a crystalline structure (Chadarevian 252). The leap from crystals formed via spontaneous order to proteins, therefore, is not actually a leap, but another seamless continuum. From crystals to crystalline proteins to simple cells to complex cells to slimes is all one smooth transition. Both crystal colonies and slimes are known to exhibit forms of mobility. "Slimes can take on various forms, and can on occasion move over the surface of other objects. In other words, they are both animate and inanimate, showing the development a rudimentary specialized tissues, behaving in ways that faintly resemble animals" (Watson 273, paraphrasing MacDougall 52). In other words, slimes are conscious.

My narrative also neglects the potential billions of molecular couplings, crystalline formations, near-RNA and near-DNA structures, and life forms that may have emerged, thrived, competed, lost, and vanished over eons. Their existence (and disappearance), like the processes I did not detail, I assumed were also implicit and known to the reader. Why we have what we have today is thanks to natural selection, even at the molecular and atomic levels. We have what we have because that is what Nature gave us.

Much current speculation and thought regarding hydrothermal vents as the sources of the origin of life treats these events as though they are static and locked in place. These locales are dynamic, not stationary; they move and shift and relocate. Thinkers along this line, like Nick Lane, however thorough his treatment, appear to be forgetting events like plate tectonics and island building. Building blocks of life formed at these locations very likely developed methods of maintaining their optimal zones and continuing to persist and evolve. In areas where plates are sub-ducted, early, crystalline, pre-nucleotide colonies of entities could have achieved a form of mobility via crystal colony growth, like we see in quartz. As plates were sub-ducted, these early pre-nucleotides could easily have crept their way up the resulting wall and remained on the ocean floor, safe from the destructive magma from which their atoms emerged. At these locales the activity of the vents would change over time. The constituent products changed as time passed and depths grew shallower. Heat energy from the hydrothermal vents would be exchanged for heat energy from the sun. Given enough time, initial amounts of hydrogen and carbon and other elements would be joined and enacted on by metals that served as catalysts for future evolution. The totality of this activity—hydrothermal vents, elementary atoms and molecules, plate tectonics, island building, crystal colony mobility, catalytic metals, and time—are more than sufficient to have formed the ingredients of nucleotides, RNA, DNA, and life.

Arthur Eddington in *The Nature of the Physical World*, published in 1928, introduced a distinction between primary and secondary laws. But primary laws drive the secondary. That is consciousness. The patterns of more complex entities is still dependent on the quantum behavior of the electron.

Speaking of my narrative, by the point at which DNA appears, we have already gone past my main argument. Let's go back to that variable electron reactive behavior—VERB—and its three results. Those electrons which hit upon the optimal state to bond with their neighbors to form nucleotides (and zircons, and crystals, and other structures that collected in the region) had "learned" a trick that would make them the "winners" in a game of natural selection. VERB, a quantum event sparked by the extremities of the environment, was both a reflex and a "choice." Electrons can adjust themselves in response to their environment. We see this demonstrated, among many experiments, in the photoelectric effect where atoms are bombarded with photons (light energy) and the electrons adjust their oscillation rates, releasing themselves from the test element, almost as if to fend off or escape the onslaught. The speed, or rate, of an electron's "orbit" around its nucleus, its trajectory, its spin rate and direction, its resonance, quantum tunneling–all of these behaviors are variable under stress. It is a classic example of decoherence–the breakdown of a quantum system not perfectly isolated, but in actual contact with its surroundings–in this case frigid ocean or molten lava, with an optimal zone in between. The onslaught of extremities is what triggered the electrons' behavior, varied and reactive (VERB), in this event on the young Earth. Deep in the oceans, an event parallel to the photoelectric effect occurred, only the "bombardment" causing the variation was the extreme heat and cold the electrons had to "choose" between. The ability to find an optimal state–resonance, "orbit", trajectory angle, spin, exchange–to enable bonding with other atoms that remained in the vicinity via the same trick—VERB—and would eventually form nucleotides through spontaneous order, was an act of behavior that had the benefit of survival. The electron's variation of its state shows an "act," and is indicative of "intent." As recently as December 5, 2018, Tam Hunt speculated in a *Scientific American* blog that electron resonance could be the basis for consciousness ("Hippies"). Clearly more than one thinker is on this particular trail. None, as of yet, have been as inclusive or expansive as my current proposal.

As I explain in section II, I am not saying an electron is a conscious being, it does not demonstrate awareness or thought. It does, however, demonstrate consciousness, its behavior is the essence of consciousness. There is a distinction. This trick of the electron is comparable to a mutation—though it is not a mutation but an inherent trait of electrons—on the genetic level that nature selected to continue, a beneficial characteristic that is fundamental to electrons (therefore a quantum event) that benefited, no, created, us all. Again, this is speculation on my part based on my understanding of the behavior of electrons, atoms, and molecules. I have no specific data to support my contention It is just a guess, but one of which I am firmly convinced. Consider, two of the three mechanisms of Darwin's theory of evolution—variation and selection—are apparent in this bonding of atoms near the ancient, sub-oceanic volcanoes and vents. The third mechanism, replication, starts to occur just one level higher on the spectrum of consciousness (discussed below). The idea that this occurrence is the foundation of consciousness, and therefore life, on Earth, to me, seems self-evident.

Jim Al-Khalili and Johnjoe McFadden in *Life on the Edge* (2014), explain the role of the electron in the process of photosynthesis. Such a vital part of life itself is dependent on a wandering electron, the "exciton" (love the name!), again demonstrating the electron's role in consciousness. Photosynthesis is an expression of consciousness. It "knows" what to do. And its behavior is rooted in the behavior of the electron—VERB. Further research has been done in several studies confirming the role of the electron's quantum behavior in multiple life processes. Speculation has been made that genetic mutation itself might be a result of quantum activity. Watson writes, "If the quantum nature of mutation is confirmed, this will help explain how diversity occurs in the first place—making evolution possible at all. Evolution in a real sense will be a quantum process" (447). He goes on to say, "If Putnam and Laughlin and Kauffman and Pregogine and Anderson and Gaylord Simpson are right, there is a major hinge in nature that occurs between elementary particles and molecules, particularly organic molecules. The latter are spontaneously organized in a way that appears to have little to do with the properties of the particles themselves. Though they do not contravene those properties, the information in the particles does affect the behavior and form of the molecules, as the quantum nature of mutation suggests" (485). Has this not been my argument from the beginning? Electron behavior is the directions, orders, instructions given to the molecules. VERB is the conductor which leads the symphony of life. Spontaneous order, mutations, natural selection, and time give us the world we see today, with the electron working, responding, and driving much of it all.

Consciousness, therefore, is not simply a result of variable electron reactive behavior; consciousness *is* variable electron reactive behavior in its simplest form. From VERB as its foundation, consciousness increases in complexity as the entities increase in complexity. The degree of complexity is both the result and the cause of consciousness. But I'm getting ahead of myself.

Oppenheim and Putnam quote Richard Benedict Goldschmidt: "The 'first complex molecules endowed with the faculty of reproducing their own kind' must have synthesized—and with them the beginning of evolution and the Darwinian sense—a few billion years ago" (24). This synthesis at the molecular level is the result of electron behavior that Bohr identified in atom formulation. One leads to the next, and it all starts with VERB spurred by the extreme environments at the thermal vents, as Goldschmidt (via Oppenheim and Putnam) says, billions of years ago.

The previous narrative, over the year since the initial publication of this paper, has led to considerable confusion for many readers. I have been asked to clarify what exactly it is I am saying occurred four billion years ago. I will try my best. The following is not new information; I'm adding nothing to the hypothesis. I'm merely slowing down and filling in the gaps that I apparently created when I published this little thought experiment in haste. My apologies.

We know from current oceanography that deep in the frigid waters of oceans today, there are many volcanic crevices and vents emitting lava, gasses, and ash. Around these vents, discovered in recent years, are anaerobic microbes and other life forms thriving in environments considered hostile to life. Also found in these areas are multitudes of other molecular compounds and formations comprised of innumerable combinations of materials spewing from deep under the Earth's crust.

The conditions found today are no different, or negligibly so, than those that existed four billion years ago. Deep oceanic volcanoes on the young Earth were spewing lava and gasses and ash back then just as they are today.

All of the makings of living and nonliving entities can be found in these volcanic emissions. My narrative focuses on the carbon atoms that were prevalent then and now. Why? Because carbon is the basis of all four nucleotides that make up RNA and DNA and therefore comprise all living things, including us.

Carbon was not the only element flittering about near these ancient underwater volcanoes. Every element found in magma was also bursting forth from underneath the crust. But thanks to natural selection, over the process of many years, carbon just happens to be the primary ingredient that led to the building blocks of life. Many other entities formed, vanished, reformed, melded with others, vanished again…who knows exactly what was happening? Not me. I know only that carbon atoms hooked up with oxygen, nitrogen, and hydrogen to form nucleotides that hooked up with one another and, thanks again to natural selection, won whatever battles they were fighting and survived to form DNA. DNA itself had the neat trick of replication. So it too won its own battles against possible other entities struggling in the region and went on to become life itself. I do not know what other molecular formations may have thrived and striven in that long ago. All I know is DNA survived and here we are today.

Backing up further to a time before the formation of DNA, we have the individual atoms in the region struggling for their own survival. The lava, ashes, and gas were too extreme for the atoms to bond and form molecules that would lead to life. The frigid waters of the ocean were too cold. Somewhere in between, at the location of the vents themselves and the surrounding ocean floor, the atoms that landed there were able to bond, survive, and form the foundation for many things including you and me. Here again, natural selection was at work. It's always at work.

Backing up again, we come to the individual electrons in each of the atoms in our narrative. We know from experimentation that electrons are able to react to stimuli of various kinds. In the photoelectric effect, electrons abandon their host metals when those metals are bombarded with photons. My argument here is that the extreme heat of the lava and the extreme cold of the ocean served as the source of "bombardment" similar to the onslaught of photons. Electrons are NOT conscious. Electrons do not THINK. Electrons do not make choices. However, electrons do REACT. And they react in a number of ways which I have already listed. They react in a VARIETY of ways (Variable Electron Reactive Behavior—VERB). Their reaction to stimuli is what determines their behavior, and therefore their home atom's interaction with other atoms. The varying behaviors exhibited by electrons in reaction to stimuli determines the fate of their atoms. In the case of the atoms at the ancient oceanic volcanoes, the electrons in carbon atoms that reacted by enabling their atoms to bond with oxygen, nitrogen, and hydrogen (again with the assistance of natural selection) went on to form life.

Now, what does this have to do with CONSCIOUSNESS, and my contention that consciousness precedes life? The behavior of the electrons is the foundation of consciousness, yet precedes actual life by far. The bonding of the atoms is a result electron behavior, yet also precedes actual life. The creation of nucleotides is the result of the bonding of certain atoms which is the result of electron behavior, yet still precedes actual life. The formation and replication of DNA is a result of the assembly of nucleotides which are the result of atoms bonding which is the result of electron behavior, yet still precedes actual life. And here we pause and reflect. Because REPLICATION requires CONSCIOUSNESS, however primitive. DNA is not thinking. DNA is not making choices. But DNA "knows" what to do to replicate itself and survive. That is consciousness. And we would not have DNA if nucleotides did not "know" how to hook up with the right partner. That is consciousness. But we would not have nucleotides if the right atoms did not "know" which other atoms to bond with. That is consciousness. And those atoms would not "know" how to bond if electrons had not reacted to the extremes of their environment four billion years ago. That is consciousness.

Photosynthetic life forms producing carbon dioxide and food consumers utilizing oxygen must have emerged simultaneously. Those life forms that had learned to create their own food were in fact a food source for those which lived on oxygen. And they lived a symbiotic existence, most likely in a limited locale, as they were the suppliers of food for one and respiratory material for both. As with my neglect to mention algorithms in the early formation of structures leading to life (see above) so here I will also not mention the millions of other possible life forms utilizing molecules other than carbon dioxide and oxygen, or consuming things for food other than each other. The focus of this paper is what we have, not what we don't have.

Picking back up from DNA, to photosynthetic molecules, to membranes, to cyanobacteria and on up the line to human beings, what we have is one long spectrum of beings and entities that span from homo sapiens at the top to single-cell organisms on down to molecules, atoms, and electrons…all of them across the spectrum exhibiting some degree of consciousness. Somewhere along the spectrum, between VERB and homo sapiens, life itself emerged. But more about that later.

Starting with complex life forms and going backwards, consciousness is exhibited in many, many entities, even at the molecular level as seen in DNA replication, mobility of crystal colonies, the encoding and proteins, and light preference of certain molecules. The semblance of intent is apparent even at this level. Discussions of consciousness, like those of Roger Penrose and associates, focusing on neuronal complexes and "microtubules" as their starting point ignore the behaviors of forms far smaller and simpler. Even these complex structures are far along on the evolutionary line of development and appear millions of years after the events which are the true origins of consciousness on Earth. No one argues that consciousness is a trait developed over time from simpler systems. My argument takes that development backwards in time along a continuum of complexity to the smallest possible level. With molecules exhibiting behavior that resembles, and eventually results in, intent, consciousness, in my estimation, must precede the molecular level and begin at the atomic or quantum level. Even ions themselves, undergoing quantum Brownian motion, are exhibiting behavior that can be reduced. My guess was that VERB was the culprit, and as far back in complexity that we could logically go.

To elaborate, we and all forms of life are not single, solid entities. Take humans alone. At the quantum level we are mostly empty space. At the macro level we are mostly water. Every single cell in our body has a mitochondria that once upon a time was an independent, single-cell organism. Our bodies are hosts of microbes, working in concert together to form our total consciousness, while also working independently each and every one. It is the symbiotic nature of all these elements working in concert together that create consciousness in all its varying degrees. And their mode of communication is electrochemical, occurring at or near the speed of light. People speak of microtubules as the foundation of consciousness, but even microtubules are the result of symbiosis of previous simpler forms. And even those simpler forms exhibited consciousness. Even those who recognize the quantum nature and behavior of particles within the microtubule and neuronal networks ignore the fact that these activities and events occur within a framework that is itself a result of consciousness of a simpler entity, not the originator of consciousness.

Having wandered afield a bit, I remind the reader that electrons are not conscious in and of themselves as entities. Their behavior is the foundation of consciousness. I know this is may be confusing as it entails a very subtle and no clear distinction. Even Einstein struggled with the idea: "I find the idea quite intolerable that an electron exposed to radiation should choose *of its own free will* not only its moment to jump off but also its direction," he despaired to Max Born in 1920. Every level of consciousness, as I shall show, can be traced back to this little trick of the electron which didn't like it too hot, or too cold, but just right, and found a way to do it. That, my friends, is consciousness. Several factors play into the behavior of electrons. I am arguing that their ability to adjust their state in response to their environment is central and fundamental to the development of consciousness on Earth. The misleading term, I realize, is "intent." I am not claiming that electrons themselves are conscious. Merely that their behavior resembles intent only in that its results–consciousness and life–are indicative of intent. It resembles intent and results in entities that actually exhibit intent. So I am declaring VERB the foundational behavior that leads to consciousness. As I discuss below, on a spectrum of consciousness comprised of ten distinct, yet interwoven, degrees or levels, "intent" does not appear until the seventh rung up the "ladder" of complexity.

Consciousness precedes life. However, life is hard on the heels of consciousness. Consciousness occurs at the quantum level. Yet life, defined over simply as exhibiting replication, locomotion, respiration, and metabolism, is quickly subsequent. Nucleobases exhibit consciousness in their assimilation and constitution in optimal environments. Their constituent parts promote and secure their survival. DNA and RNA exhibit consciousness in their intentional arrangement, also in the fact that they replicate themselves. Yet they are still not life. At which point life itself begins I will address briefly below.

The origin of consciousness on Earth, and therefore the origin of life, could only occur in an environment provided by deep ocean volcanoes and volcanic crevices. In the surface atmosphere or in a gas environment, the atoms would scatter; in a solid, rocky environment they lack the necessary mobility to meet up; in a molten environment they break apart and do not bond. In a liquid environment of extreme temperatures, however, in between the extremities, an optimal zone exists. The ability to maintain position in the "Goldilocks" zone between the furnace and frigidity is the epitome of consciousness.

This is not to say that life could occur only on Earth. Such liquid environments are not exclusive to our planet. Other celestial bodies, including exoplanets, moons, meteors, and comets, have similar conditions and could host the same event.

In regards to why consciousness, and life, on Earth is carbon based as opposed to silicon based, it could very well be due to the relative weights of carbon and silicon and the consequences of this relative weight in this narrative. Though they could be likely partners and seem to get along just fine in laboratory experiments, in the extreme sub-oceanic environments, they are too far apart on the periodic table to find one another and bond in the optimal zone. Silicon, though abundant in magma, would tend to settle deeper than the three next-door neighbors—carbon, nitrogen, and oxygen. Silicon was more likely to remain in the magma below the crust. This is, of course, just more speculation.

The carbonic molecules formed as a result of VERB and subsequent bonding at the sub oceanic volcanoes and crevices were heavy enough to remain collected in the optimal zone and may well have formed vents similar to ones we see today. The maintenance of order in a system requires considerable energy. Maintaining equilibrium in an optimal zone, and thus survival of the atoms and molecules in a state capable of forming early carbonic molecules, is a form of order. The energy utilized here was initially the heat of the lava, ash, and gasses. As waters grew shallower as a result of plate tectonics, volcanic island building, and crystalline "mobility" (symbiotic colony growth being an early form of mobility), some molecules exchanged heat energy for sunlight in an event that eventually became photosynthesis. Lower (as in depth) molecules soon began to reap the benefits of light energy through chemical connections that form the earliest ingredients of "life" as we know it. Those molecules which form the nucleobases form colonies which eventually "learn" to photosynthesize, leading to life, albeit post-consciousness. Consciousness is already a thing by this point. The attraction and beneficial paring of atoms and molecules is at this point "intentional," as well as in keeping with the idea of spontaneous order. The electrons (and their atoms) are "happy." We are well past the appearance of consciousness and well into the formation of life by this point in the narrative. And, yes, again this is my own personal conjecture.

When the temperature of metals is reduced to a critical level, superconductivity appears, whereby the electrical resistance totally vanishes and electrical currents, once started, persist indefinitely. Whole technologies have been built on these phenomena but their significance here lies in the fact that they show spontaneous order (Taylor 290-295, cited in Watson 434). This is what we see at the thermal vents and why I argue they are most likely the source of the origin of consciousness and of life. Because of spontaneous order and the direct link to temperature, the appearance (emergence) of complex molecules at the vents could have been very rapid.

And now for the crème de la crème of Section I. The following list and definitions of the levels of consciousness I initially jotted down in a note that took approximately ten minutes to write. Each term and a one-sentence definition seemingly spontaneously generated in my brain. I set myself the task of listing and defining the levels of consciousness with not a single preconception of what the hell I was about to write. I composed the list at a very early stage, when the idea of a spectrum of consciousness was barely a thought, and the concept of and term VERB had yet to find expression. The initial list of terms and definitions was part of a handful of notes—fourteen pages in total—that I would eventually flesh out into the 32-page publication I posted on November 15, 2016.

Consciousness is expressed on a spectrum. There is a correlation between complexity of an entity and the degree of consciousness it exhibits. In general, as molecules, and eventually the life forms they are the basis of, grow in complexity, so does the degree of consciousness.

The spectrum of consciousness as it exists on Earth, in both pre-life and living entities, includes, but may not be limited to, the following ten degrees: Reactivity, Bonding, Instinct, Essential Awareness,

Cognition, Permanence, Intention, Foresight, Self-Awareness, and Global Awareness. Between Bonding and Instinct comes the actual beginning of life, though Reactivity and Bonding are both still degrees and forms of consciousness. Each degree above Reactivity on the spectrum of consciousness also exists as a spectrum, each more broad than the next. Each degree of consciousness is dependent on the one which precedes it.

The argument may be made that Reactivity and Bonding are not expressions of consciousness, but this would be an error. Variable electron reactive behavior (VERB) is consciousness, for thought itself, experienced at the macrocosmic level, is merely a series of electrical events—electrons in their dance. Thought and its effects is simply the electron dance in a more complex system.

I will briefly define each degree of the spectrum of consciousness. Granted, the following terms and definitions may borrow terms already in use in other fields of psychology and theories of mind. The definitions, however, are wholly my own.

A friendly reminder: At every point along the following spectrum, natural selection is at play.

REACTIVITY: Reactivity is the simplest degree or level of consciousness. It is the variable electron reactive behavior (VERB) occurring at the quantum level. When an electron reacts to a change in its environment, it does so in multiple ways: by adjusting the speed of its orbit around its nucleus, or its trajectory, the rate or direction of its spin, or even may pop out of and back into existence altogether, in probabilistic manners. Reactivity is the level of consciousness upon which all others are built. There is no degree of consciousness on Earth simpler than Reactivity. Regarding natural selection, all electrons react, causing their respective atoms to behave in various ways. Some do nothing of significance. Others find their way to the next level.

BONDING: Bonding is the next level and is dependent upon Reactivity. Bonding is the degree of consciousness where individual atoms bond with others to form molecules, where molecules bond with others to form membranes and bases, these form networks, and so on up the chain of complexity. Again, bonding occurs in all kinds of ways. Only those molecules selected by nature make it to the next level of complexity and bring us closer to the world we know today. Here we find the periodic bonding of salt and quartz crystals, precursors of the aperiodic DNA "crystals" that appear in the upper end of this level and lead curiously and seamlessly into the next.

INSTINCT: The next degree of consciousness, Instinct, is the level where we first observe actual life. Instinct is not exclusive to life. The movement and sprawl of certain crystal colonies that are non-living yet exhibit characteristics of life are an example of Instinct in non-living entities. The tendency to seek out optimal environments in which to thrive, even though they, the crystal colonies, are non-living, is possibly a wonderful example of consciousness at this level. Even more wonderful, replication of DNA strands, not simply bonding, but "knowing" how and into which structure to bond, is a low-level, though still quite impressive, example of instinct. Instinct, of course, is observable in all living entities. For arguments of quantum processes involved in multi-cellular life forms' (including birds and fish and humans) expressions of what we hitherto have labeled "instinct" see *Life on the Edge* by McFadden and Al-Khalili.

ESSENTIAL AWARENESS: This degree of consciousness is the earliest one occurring on the spectrum where true awareness becomes detectable. We see this in the awareness of a plant knowing where the sun is located (heliotropism). We see this in a sperm cell's awareness of where to find the ovum. We would certainly not call a phagocyte a thinking being, yet it is obviously aware that a bacterium is in the vicinity and is able to act accordingly by gobbling it up. These are all examples of Essential Awareness.

COGNITION: Cognition is the earliest degree of consciousness on the spectrum where a "thinking center" is required—a brain. Though this degree also occurs on a spectrum, as do all degrees of consciousness, this is the earliest level where we can confidently say that thought is taking place. The emergence of thinking centers (brains) on Earth, of course, are another result of natural selection. The ability to think is clearly advantageous to survival over not thinking. The spectrum of types of thinking centers–brains–from very primitive, yet functional, to the majesty of the human brain is a result of evolution through natural selection. Those of us at the "top" of the complexity chain are simply "winners" in a billions-of-years process. Although natural selection continues today as a determinant in evolution, with the emergence of human brains and, consequently, of culture, natural selection plays a less decisive role in the evolution of human consciousness. It is here that language and memes and culture itself and technology take an upper hand in the development of the human mind (if such a thing actually exists).

PERMANENCE: Permanence is the degree of consciousness on the spectrum where beings are aware that entities interacting with them and sharing their environment do not cease to exist when they are not in their line of sight or range of hearing.

INTENTION: Intention is the earliest degree of consciousness on the spectrum where choice becomes evident. A cognitive entity, aware of the permanence of the entities and objects around it is able to make a decision regarding them; options become apparent to them and they are able to choose from among them a course of action. Though uncertain or unaware of the possible consequences, the entity is able to think, though not necessarily with actual words, but in essence or the equivalent of, "Between A and B, I will do B."

FORESIGHT: At this degree of consciousness, a being is not only able to make a choice between two or more courses, the being is able to predict possible outcomes. An entity exhibiting the degree of foresight is aware of consequences, is cognitive of cause and effect, and can act or choose not to act on a course of action as a result of reward or punishment.

SELF-AWARENESS: It is at the degree or level of Self-Awareness that an entity becomes aware of itself as a being distinct and separate from other beings with which it interacts. This is the level where an entity, though it may not possess language, has the concept of "I." As with all levels of the spectrum, Self-Awareness exists on a broad spectrum. From the basic awareness of separateness as a being from other beings and the recognition of an "I," to self-examination and the varying degrees of intellectual activity, the spectrum of Self-Awareness is vast. The standard definition of consciousness is really what I identify as Self-Awareness, a level of consciousness occurring next to last in increased complexity along the spectrum. Self-Awareness is consciousness, yes, but it is a highly evolved expression of a phenomenon that we can trace back to the quantum level. Having a sense of "I" is not the be all and end all of consciousness. It is neither the starting point, nor the culmination. It is merely one, albeit a highly complex, expression of consciousness. Much has been written and discussed about "I" and "I-ness" as an emergent property of the amalgamation of multitudinous, many unknown, phenomena and processes. This paper strives to trace the origin of all those processes back through the levels of complexity and finally pinpoint the origin of consciousness in a quantum event.

GLOBAL AWARENESS: Having built upon all previous levels of consciousness on the spectrum, a being that achieves this level (and to my current knowledge only humans can be said to have done so, though an argument can be made that whales and dolphins have also achieved this level), becomes aware of its relationship and impact on all other entities with which it interacts, as well as those it interacts with only by proxy or through a chain of events, and those with whom it does not interact but is aware of their existence, even to the whole Earth and beyond. At this level, a being is not only conscious of an other, it is aware of all others, and not just those of its own species. All humans, all animals, all plant life, the awareness of species lost to extinction and the possibility of species yet to be discovered—these are all thoughts which can occur to a being at this degree of consciousness.

Though the shorter title of my paper is "A Quantum Origin of Consciousness," at no point do I mean to imply that a brain, not even a human brain, is a quantum machine. Charles Seife makes very clear, despite some believers and mystics, the brain is classical in its operations (215). What Seife and others have not explicitly considered or written about is the spectrum itself. The appearance of brains and even lesser entities is far down the evolutionary line from my initial proposal: consciousness itself is the result of quantum activity—VERB.

Researchers still struggling with the "hard problem" of consciousness—locating the "seat" of the "I" in the brain—are missing the point. Consciousness is the product of our total being, our brains and our bodies and our environment and (for humans) our capacity for language and our enculturation, all of which are resultant of our complexity, our location along the spectrum of consciousness. Even species lower on the spectrum exhibit some conception of an "I" while some humans, due to developmental disabilities or traumatic brain injuries, do not.

Oppenheim and Putnam also enlist Ludwig von Bertalanffy: "Reality, in the modern conception, appears as a tremendous hierarchical order of organized entities, leading in a super position of many levels, from physical and chemical to biological and sociological systems" (29). What Bertalanffy was describing, quite succinctly, is the spectrum of consciousness, extended, at last, into the sociological realm.

George Gaylord Simpson, in *This View of Life* and other papers, also addressed the spectrum of consciousness, though not in these exact terms. He called it the "evolutionary intensification of awareness" by which we know the mechanisms that give us the ability to know them (Watson to 41).

The spectrum of consciousness varies even within a species. One mammal may act purely on instinct while another mammal of the same species may exhibit foresight, as may be seen in the use of tools by some species, like using sticks for fishing or to gather termites from a mound (Goodall 12).

If, as I argue, consciousness occurs at the quantum, atomic, and molecular levels, then consciousness precedes any complex "machines" natural selection and time can produce. Now, the higher levels of consciousness, like cats and dogs and elephants and dolphins and humans, are results of Nature's ability to build complex machines. "Efficient" is a term I'm not comfortable with. Some schools of thought could point to the possibility that Nature's goal is to manufacture "machines" and beings capable of comprehending her. Maybe even "worshipping" her. At least communing with her. I'm not actually saying that is the case. But it's a nice philosophical way of looking at Nature as possibly conscious herself. I am not a panpsychist. But I don't rule it out.

Where people often go astray is that they have ONE idea of consciousness. It is or it ain't. I'm proposing consciousness is actually a very, very, VERY basic and simple phenomenon at the lowest scales that, as complexity of entities increases, so does the complexity of the experience of consciousness. Culminating, perhaps, in human consciousness...at least it is the highest I am aware of.

I was recently challenged by a colleague regarding the viability of my hypothesis as being "scientific." To be "science" it must be tested. "How, exactly, could you test your idea?"

Glad you asked: Reproduce the conditions of the thermal vents and see if the constituent materials follow the pattern I've predicted. Monitor electron behavior and see if it influences the formation of pre-nucleotide molecules or no. It would not be possible in real time, but computer models could predict the processes accurately enough to confirm or reject the path I say consciousness and life took. See if the extreme heat and cold of the environment leads to formation of pre-nucleotide molecules in an optimal zone. Once the process is started, it follows a clear path toward living beings. Just need to see if what I say started the process is actually capable of starting it or not. The exact opposite could be true. The conditions I say started the process may actually be counterproductive to it and eliminate it as a possibility.

I've given reminders that evolution through natural selection has been key in every level of complexity along the spectrum of consciousness. Increased complexity is the key to survival. Increased complexity is the result of "winning" the selection battle, as well as the key to survival itself. Much of this is dependent upon the "good tricks" that each entity and species develop. Good tricks are not limited to only to lower animals acting primarily on instinct. "Good tricks" appear all the way up the spectrum of consciousness and become more complex up the higher they emerge. One of humankind's first good tricks was bipedalism. Another was standing erect. Our next good trick was the development of language. The discovery and development of technology is another good trick. Writing is also a good trick. Each of these developed to the increase and expansion of our consciousness as a species, as well as the acceleration of our evolution and development of culture, to the point that language and culture expanded our consciousness exponentially as biological evolution (though still active) took a less obvious role. Many of our species point to human capabilities like language and culture as hallmarks of our specialness in nature. When in reality, like all animal species, we just have some really good "good tricks."

From the simplest level, Reactivity (VERB), to the most advanced levels of Global Awareness, all of consciousness is still essentially response to stimuli.

However, at the uppermost degrees, what constitutes stimuli expands from heat or cold, hunger or thirst, physical injury or sensual gratification, to things like dreams, memories, fears, and emotional injuries. Thought itself becomes stimulus for more thought.

Again, consciousness varies in degree. On Earth, it exists on a spectrum. People yearn for a simple definition, "What is consciousness?" And there is no one answer. It is different for each entity who experiences it. If I'm forced to give a single, honest definition, I would have to say "Consciousness is the totality of experience available to an organism based on the tools allotted to it by evolution." But that is oversimplifying it. By breaking it up into degrees and levels, I am defining what consciousness is for those entities at each level. What is consciousness as experienced by an entity at one level is different from consciousness as experienced at a lower or higher level. Yet, overall, from the electron reacting to stimuli, to the brightest of humans solving world problems or mysteries of the universe, *all* of it is consciousness to one degree or another.

One may argue that these degrees or levels of consciousness are not only arbitrary, but bogus. How does one distinguish one level from the other when the behavior of entities within two adjacent degrees may be indistinguishable? What really is the difference between Instinct and Essential Awareness? Are they not the same? This is a valid argument, considering that the spectrum of consciousness exists on a seamless continuum. The spectrum is not a list of ten distinguishable levels or degrees, but an unbroken path from the simplest to the most complex. So, between the uppermost expressions of Bonding and the lowermost of Instinct, there are no distinctions. Where does bonding to form crystals become crystal colonies that capitalize on their environment and exhibit an early form of locomotion in a seeming search for the optimal living space? One blends seamlessly into the next. Between the upper most examples of Instinct and the lowermost of Essential Awareness, there are no distinctions. Crystal colonies seeking optimal living environments are not aware of other entities around them in the same way a phagocyte is aware of a bacterium nearby. One degree meshes into the next. And so on up the spectrum. The degree of Bonding, for instance, includes everything from the formation of the simplest molecules, like water, to the formation of DNA strands. However, by the time we reach the upper end of the spectrum of Bonding (DNA strands with their replication, crystal colonies with their apparent mobility), we have crossed seamlessly into the lower end of the degree of Instinct. Exactly where one becomes the other is not distinguishable. Even between the highest expression of Self-Awareness and the most basic expression of Interconnectivity, or Global Awareness, there is no distinction. Even at the lowest level of Self-Awareness, an entity is able to distinguish "I" from "Other." At what point does that entity become aware of all others? Somewhere along the continuum, but at no certain or identifiable point.

In *The Structure of Science*, published in 1961, Ernest Nagel introduced the doctrine of emergence. "It is of the essence of Emergent Evolution that nothing new is added from without, that "emergence" is the consequence of new kinds of relatedness between existents" (372). It is the complexity itself, the result of natural selection, "winning," that enables the "emergent" behaviors and good tricks we see in species, including humans.

Philip Warren Anderson elaborates on this idea of emergence and what he refers to as "constructivism," saying that in nature rules—laws—of complexity that while they never break the laws of particle physics add to them new laws of construction which are every bit as fundamental as the simpler laws governing electrons of photons, for example. As we go on up the hierarchy of sciences, "We expect to encounter…very fundamental questions at each stage and fitting together less complicated pieces into the more complicated system and understanding the basically new types of behavior that can result" (396, cited in Watson 433). This is in essence a description of the spectrum of consciousness and how emergence of behaviors and complexity of consciousness increases as we ascend.

Not only does consciousness appear on a continuous spectrum, and not only is each degree built and dependent upon those preceding it. We can think of consciousness like a combustion engine, with each degree being equivalent to a cylinder. The degree of Reactivity is a one-cylinder engine; Bonding is a two-cylinder engine. Global Awareness is a ten-cylinder engine. At every level, all cylinders of that particular degree are "firing" all at once. Not at any time does activity occur on any given level without all other levels being engaged. Not to promote or encourage militarism, but a great example of a ten-cylinder engine–Global Awareness–would be a military leader plotting the retaking of a city controlled by, oh, by whoever he and his bosses don't like. His decisions include the enemy, the civilians, collateral damage, political, social, and economic outcomes of a victory or a defeat, promotions or demotions, his own personal conscience, the cost in both human and monetary terms, and on and on. This individual is firing on all ten cylinders. From the awareness of the other persons involved, known personnel and unknown civilians, all the way down to the electrons dancing in his brain with each consideration, every degree of consciousness is active in such an event.

I would be remiss if I did not mention the role of language in regard to consciousness. Some would argue that language is essential, if not necessary, to consciousness. Clearly, I would say they are wrong. Language emerges on the upper end of Self-Awareness, and is pervasive (and perhaps, necessary) to Global

Awareness. Language may also be, and most likely is, responsible for the leap in the expression of consciousness experienced by human beings. Yet, I would argue that, rather than being a requirement for consciousness (it is not), it is, however, crucial to the degree of consciousness one achieves. I would also contend that language becomes even more important as one reaches the upper end (whatever its penultimate point may be) of Global Awareness. In other words, it is arguable that the more advanced our language skills (as individuals), the more advanced we may be along the scale. This is not to say an individual's vocabulary solely determines how "far" they can go. Persons with certain developmental and social disorders may have astronomical vocabularies, and yet be on the cusp or even shy of self-awareness. Hindu yogis, Buddhist monks, and mystics of varying traditions may have comparably small vocabularies, and yet exhibit or possess a consciousness bordering on universal or cosmic. I know I risk slipping into metaphysics with such a remark, but who's to say such a degree of consciousness is unreal and unattainable?

Clifford Geertz addresses the idea of co-evolution (biological and cultural) in his two master works, *The Interpretation of Cultures* (1973) and *Local Knowledge* (1983). Watson paraphrases Geertz's argument: "It is wrong in his view to assume that the brain of Homo sapiens evolved biologically and that cultural evolution followed. Surely, he argues, there would have been a period of overlap, of co-evolution. As humans develop fire and tools, our brains would have still been evolving—and have evolved to take into account fire and tools" (308).

One thing that distinguishes beings in the upper ranges of the spectrum versus those in the lower ranges is the capacity for memory and reflection. As regards the spectrum of consciousness, all living creatures (beings) and even some non-living, or pre-living, beings have the rapid automatic thoughts (let's call them "responses"). Capacity for the second kind—slow, contemplative, reflective thought—increases as complexity increases higher up the spectrum.

Then one asks the questions: What are memories? What are dreams? We know many species other than humans experience both memory and dreams. What are they, in reference to consciousness? Which is like asking, still, WHAT is consciousness? Memories and dreams are the same as all sensory experiences, which for mammals, at least, consist of five (arguably more). Our experiences of sight, hearing, taste, touch, and smell are all a result of the interplay of our nerve endings (which receive them) and our brain (which "interprets" them). Memories (intentionally or unintentionally recalled) and dreams (definitely unintentional) are merely our brain doing what it does in regard to our sensory experiences MINUS the senses. It is the brain doing the same job it does in conjunction with our nerve endings, only without the nerves. And the biochemical response to these memories and dreams is equal (or even more intense) to that of actual sensory experience. Our brain can make us experience joy or sadness or fear or pain, even while we are alone, staring idly out a window, or even while slumbering.

Admittedly, human consciousness is far advanced in comparison to that of any other animal. Yet, it still is only a matter of degree, not a matter of essence. Yes, our brains are larger and have far more connections than those of our closest relatives, chimpanzees and bonobos. But there is nothing other than size and number of connections to distinguish a human brain from that of a bonobo. We have not evolved or inherited a separate, identifiable "thinking center" in our brain that bonobos lack. Many factors other than simply brain size have led to the exponential acceleration of consciousness as experienced by humans: bipedalism, our meat-rich diet, and, of course, language and culture. So what is consciousness for humans is indeed different from what is consciousness for other primates. But that does not mean we can draw a clear line between our consciousness and theirs. No more so than we can draw a line between the consciousness of an adult human and an infant human. The mechanisms for consciousness in both are genetically identical. The difference is one has acquired language and inherited a culture. The other has yet to do so, but will, given enough time and good luck.

Consciousness, though rare, is fundamental to the universe. This is not to say I promote or believe in panpsychism, which I do not. I would sooner say voodoo is the source of consciousness than to suggest panpsychism. I am a materialist and a reductionist. I merely mean the ingredients for consciousness can be found many places in the universe, and that it is not too complex of an event to occur in more places than one. Its appearance is by no means limited to Earth. Nor is it restricted to planets with liquid water. There must be, however, some liquid for the constituent parts to bond. Water/liquid is essential to the formation of the structures of consciousness and of life. Rosalind Franklin's criticism of Watson and Crick's early assessment of the structure of DNA was that there initial model did not "take any account of the fact that in nature DNA existed in association with water, which had a marked effect on its structure" (Watson 229). It is for this reason that water, or some liquid element, is necessary to the foundation of life. There must also be environments of extremities, spurring the localization (gathering) of the constituent parts and therefore encouraging the bonding that leads eventually, and seemingly inevitably, to life. We are most likely not alone in the universe.

We could not predict a phenomenon like consciousness or life, but it's pervasiveness, and it's relative ease of construction, make it in evitable. Weird. Inevitable, but non-predictable! Case in point: a coastline is a living thing, ever-changing, as a result of multiple events. Though we know the processes that form a coastline, we could never predict the shape of coastline will take. Chaos out of order: as complexity increases, even though at it's root this equilibrium, the final product (a human being) is holy unique and non-homogeneous.

So the question remains: Where and when did life begin?

When we consider the spectrum of consciousness, essentially there is no such thing as an origin of life. Consciousness was already established at the quantum level, and as complexity of molecules went from bonding to instinct, which we see expressed in crystal colonies, to DNA replication, to photosynthetic molecules, to cyanobacteria and on, life already was. Life is just the natural result of increased complexity along the spectrum of consciousness. Life did not begin. There were only the degrees of consciousness along a spectrum of increasing complexity of the entities involved. We see this exhibited in Nature today where we have entities that are clearly non-living, entities that are fully alive, and some, like viruses, that are somewhere in between. The complexity of the organism is what determines the degree of consciousness. But being a continuum, there is no point at which life originated.

Consciousness is not a trait of life; life is a trait of consciousness.

In his 2018 book, *When Einstein Walked with Godel*, Jim Holt suggests that one of the most pressing human questions regarding nature--"how the brain gives rise to consciousness"--may remain unanswered. Even this late scientists and philosophers and thinkers of all bents, like Holt, are still missing the point, getting it backwards. The brain did not give rise to consciousness. Eventually, and, perhaps, inevitably, consciousness gave rise to the brain.

Life *is* consciousness, only a few steps up the spectrum. Researchers who refer to the beginning of life or the "spark" that lead to living cells are on the wrong path. The spark was there all along at the quantum level in the form of VERB. From electrons in individual atoms shifting their behavior in response to stimuli, through the formation of molecules, crystals, and membranes, through the replication of DNA, to you and me, there is a seamless spectrum, all built upon the previous level. As far as "life" goes, there is no "begins."

If my hypothesis is correct then, not only did life not have a specific beginning, it continues to begin over and over and over. A cornucopia of new life forms could be emerging each day. Rather, a collection of forms that are already approaching life find themselves in an environment conducive to successful replication and, through variation, eventually become an organism we recognize as living. And if they thrive long enough and evolve far enough, they may eventually become sentient, even self-aware. The seed of the dominant life form on Earth four billion years from now may be swimming at this moment near an oceanic thermal vent, just like our ancestors were four billion years ago.

II. VARIABLE ELECTRON REACTIVE BEHAVIOR (VERB)

Dear reader, prepare thyself for snark.

In Section I, I threw about the phrase "variable electron reactive behavior" (VERB) many times while only briefly alluding to what it signifies or means. That term has cause me considerable grief in that a large section of my readers, especially early on, took issue with me coining a term. As if coining a term, identifying a thing as something new in reference to a phenomenon—in this case, electron behavior in reference to consciousness—were something wholly unheard of, if not taboo. It is perhaps my coining of the phrase "happy electrons," as opposed to VERB itself that may have ruffled some readers' plumage.

Allow me to me clarify.

VERB is not a new discovery based on new research and data. It is merely a new term to describe already-established behavior of electrons. VERB is simply my term for electron behavior as it relates to consciousness.

Electrons "orbit" the nuclei of their atoms in a quantum cloud of uncertainty. I realize that "orbit" is a troublesome term when considering the probability factors of actual electron behavior. An electron's orbit is not comparable to a planet's orbit around a star. Their directions may be elliptical, or angular, or direct, or they may pop in and out of existence entirely. But they are never stationary. Regardless of their path, electrons do revolve around their nuclei, albeit in strange and non-conventional manners, and the speed and direction of that path is variable, dependent on the conditions of their environment. The electron, in reaction to its environment, is able to adjust its speed, change its direction, increase or decrease its resonance, pop in and out of existence, speed up, slow down, and reverse its spin–pretty much anything the electron wants to do. This ability to adjust its behavior is critical in an electron's life, and therefore critical to its atom, the molecules it comprises, and all of us as well.

This indicates that "selection" occurs not only on the genetic level. Selection and "variation" occur at the subatomic level in the behavior of electrons that "choose" their state and subsequent paring partners for advantages in the extreme environments. The molecules the atoms comprise survive to replicate themselves through continued "successful" paring, on up the chain of complexity.

VERB occurs and has the same beneficial results in all environments occurring post-consciousness. I.e., VERB at sub ocean volcanoes is consistent with electron behavior in organic environments, including our own bodies. Adjusting its behavior to allow atoms to bond with other atoms at the deep ocean volcanic events is the same event as the electron activity in our brains that make us thirsty when our water volume dips below a level that makes the electrons "happy." It is the exact same electron activity in the brain that is preconscious and triggers events like perspiration, salivation, and emotions, maintaining and creating its own optimal zone within the bodies of humans, all living things, and non-living things all the way to the bottom of the spectrum of consciousness. All of these events exhibit apparent intention on the part of the electron that are pre-cognitive and occur at the quantum level at or near the speed of light.

VERB determines so many things. Its resulting inertia determines the relative "heaviness"–not to be confused with its atomic weight–of an atom.
VERB determines whether or not an electron exchange occurs. Its variability is a reaction to its environment and its ability to adjust itself indicates intention and exhibits the Will to Equilibrium (more on this in Section IV), thus making VERB consciousness. From the quantum level to

the macro, everything is determined by the behavior of the electron. Our thoughts and our emotions, our desires, our physical needs, thirst, hunger—these are all electrons seeking equilibrium, not just for themselves but for the molecules, the cells, the membranes, and the living creatures that they comprise. Quenching thirst is an electron balancing out the water level of our bodies. Hunger is an electron seeking fuel for the electrical activity of one's body. Being sick and running a fever is electrons attempting to rid the body of what ails it and establish equilibrium.

VERB is the result of electrons seeking equilibrium in extreme environments, their attempt to avoid negative stimuli or to seek positive stimuli. Though electrons themselves are NOT conscious, VERB is a reaction, a reflex, and is the foundation of consciousness as it exists on Earth.

VERB *is* consciousness, albeit the simplest, most-irreducible form. It is where consciousness on Earth begins.

I am not saying electrons are *aware*. This has been a point of confusion from the start. The trouble many readers are having with this concept, and I fault myself for lack of clarity in my attempt to explain it, lies in making the distinction between "consciousness" and "conscious beings." Consciousness and being conscious are two different things. Being conscious, or aware, is a degree of consciousness far along the spectrum as it exists on Earth. As with any spectrum, there are points were the argument for being conscious is speculative, while exhibiting consciousness is not. Is a plant conscious or aware of the location of the sun in the sky? Maybe, maybe not. Does it exhibit consciousness in the turning of its leaves toward the sun (heliotropism)? Absolutely. It is by this line of logic that I argue electrons are not conscious, but VERB is an expression of consciousness, to repeat, the simplest, most-irreducible expression of consciousness on Earth. In the photo electric effect electrons abandon their parent metal when bombarded with photons. Electrons do not do this randomly. They are perfectly satisfied where they are until the photon bombardment begins. They do not "like" this bombardment, and they do whatever necessary to get out of there! What is this but a reaction to negative stimuli? And what is consciousness itself except a response, however complex, to stimuli?

The phrase "variable electron reactive behavior" may seem redundant, or over-wordy. But each word has its place. "Variable" indicates that the options, or "choices" an electron can make regarding its state are numerous. The electron may elect one state or another at a whim depending on its environment. "Electron" because that is the quantum particle we are dealing with in regards to consciousness on Earth. Electrons are the smallest particles known to exist in isolation on our home planet. "Reactive" because the changes in state exhibited by electrons are in "reaction" to volatile conditions, bombardment, and the need to seek an optimal zone between extreme environments. I emphasize volatile conditions in reference to the origins of consciousness on the early Earth. In reality, electrons react to all stimuli, regardless of volatility. And, finally, "behavior" because that is what we observe. The electron is *doing* something in response to its environment, just like every structure, entity, and life form created on up the ladder of complexity behaves in response to their respective environments.

How important is VERB? I cannot over stress this point. By giving VERB such a brief treatment, I fear I may mislead readers as to how vital is it is to all degrees and experiences of consciousness. I published this paper over a year ago. Yet only recently (November 19, 2017), I came across this in Daniel Dennett's most recent book: "We won't have a complete science of consciousness until we can align our manifest-image identifications of mental states by their contents with scientific-image identifications of the subpersonal information structures and events that are causally responsible for generating the details of the user-illusion we take ourselves to operate in." I submit here that VERB is quite likely the phenomenon that aligns the manifest- and scientific-identifications, and could very well be the clue to a complete science of consciousness. We shall see.

Finally, as one critic put it, "VERB? That's not even a thing. You made that up!" To which I responded, "Exactly!" To that point, if Google had been in existence in 1904 and someone had searched for "special theory of relativity," their search would have come up empty. Einstein coined the term to describe phenomena he observed in reality. Likewise, all of the behaviors exhibited by electrons and accounted for by "VERB" have been observed, recorded, and even utilized to human purposes. In coining "variable electron reactive behavior (VERB)," I was merely seeking a term that was all inclusive--one phrase to describe ALL behaviors of electrons as they relate to consciousness on Earth. So, yes, prior to November 15, 2016, VERB was NOT a "thing." Since that date, however, it is. Deal with it.

III. 0=1

And now we arrive at a very problematic section. Many readers who make it this far abandon the endeavor when they read the section heading. "0=1? What the hell? We thought this guy was bonkers with the whole 'consciousness precedes life' nonsense. Now this?"

Yes. Now, this.

Let me begin by saying 0=1 is a false statement. Mathematically speaking, 0=1 is an untrue equation. It is impossible. It always has been and always will be. I hope that settles a few minds.

That being said, however, hear me out. In mathematics, 0=1 is not true. It's a false equation that is mocked and ridiculed by anyone with even the most tenuous grasp on basic math.

But Nature, however eloquently described by mathematics, is not a mathematician. Time and again in the history of science, she has proven to have secrets we humans have taken millennia to crack. Nature, if I am correct, has yet one more secret.

In my musings regarding consciousness and the behavior of electrons, I was compelled to consider the following scenario culled from previous reading in multiple studies: A man moves his right arm. Prior to that he has the conscious thought, "I will move my right arm." Prior to that is the intention of moving his right arm, an intention that he is not aware of because it is a precognitive event, yet one that can be measured by fMRI—the electrical activity (i.e. the electrons in his brain), are detectable. This has been demonstrated by researchers monitoring the brain activity of a tetraplegic patient. Sam Harris, in his book *The Moral Landscape*, and Peter Watson, in *Convergence*, both reference this wonder of modern technology. The details appeared in Science in May 2015 (Aflalo). This discovery led me to the question, "What precedes the intention?" And the answer is nothing. As regards electron activity in the brain related to the intention, nothing precedes the intention. So, arbitrarily, I gave the intention the mathematical equivalent number 1—in binary language, "on." I gave that which preceded the intention the mathematical number 0—binary for "off"—because nothing precedes the intention. Logically, therefore, it became evident to me that there must be a non-decimal point "between" 0 and 1.

There is also this, from a book published in 2018. In WHEN EINSTEIN WALKED WITH GÖDEL: EXCURSIONS TO THE EDGE OF THOUGHT by Jim Holt, the author discusses the "avatars" of higher mathematics proposed by Alexander Grothendieck. The author then cites Michael Harris and his description of a ladder of such avatars, each attaining a higher and higher mathematical truth. The author continues, "And what lies at the top of this ladder? Perhaps, Harris suggests with playful seriousness, there is "One Big Theorem" from which all other mathematics ultimately flows — "something on the order of samsara=nirvana. Every veil lifted reveals another veil." Jim Holt goes on to say, and I quote at some length: "Thanks to Gödel's second incompleteness theorem -– the one that says, roughly, that mathematics can never prove its own consistency — mathematicians can't be fully confident that the axioms underlying their enterprise do not harbor an as yet undiscovered logical contradiction. Indeed the discovery of such an inconsistency would be fatal to pure mathematics, at least as we know it today. The distinction between truth and falsehood would be breached, the ladder of avatars would come crashing down, and the One Big Theorem would take a truly terrible form: 0=1."

The time of the One Big Theorem has come. And it has taken its "truly terrible form": 0=1

The history of science is rife with moments when humankind had to revise its conception and understanding of reality. Newton, Darwin, Einstein, and Hawking were just a few of the harbingers of those moments. I am not comparing myself to Newton or Darwin, nor saying I am on the same level as an Einstein or a Hawking. Okay, yes, I am. But that little megalomaniacal quirk aside, I have a certainty, a conviction, of the rightness of my ideas that seems to accompany newly discovered truths (a certainty and conviction shared by Newton, and Darwin, and Einstein, and Hawking in regard to their conceptions). But enough with delusions of grandeur.

If I am correct, and the rest of this paper will attempt to convince the reader I am, then we indeed have reached one of those moments in history when we must revise our understanding of how the universe works, what Nature is truly capable of doing. And it may run counter to the ingrained belief that mathematics are not only critical to our description of the universe, but supreme, even exclusive. We may have to admit that mathematics are not sufficient. Nature may go one step beyond mathematics.

What exactly do I mean by 0=1, if, as I have admitted, it is an untrue statement? Only that 0=1, as an aspect of reality as yet undiscovered, is the fundamental expression of what Nature is able to do. It is not an actual mathematical equation. It is more of a metaphor for what Nature is doing. 0=1 represents a point/moment so small and so brief as to comprise zero dimension and zero duration, occurring at all points, at all times, everywhere and everywhen, forming a seamless field comprising the entire universe, the fabric of existence itself. Where positive charges and negative charges are equal. The center of the right-pointing arrow in chemical equations. The moment when Schrodinger's cat is both dead and alive. The threshold where Nothing and Everything, zero and infinity, meet.

This field, this canvas on which reality is painted, I have named the O-field. The complete theory including my description of the 0=1 point/moment, its relation to the Big Bang, the properties of the O-field, and the myriad other physical phenomena which my hypothesis helps to explain, I call O-Theory. But I'm getting ahead of myself again.

This explanation, at this point, seems over simplistic, and I have already received unsatisfactory "explanations" involving red and green lights and superposition and others. With no logical refutation beyond, "That's impossible," I decided to "accept" the idea of 0=1 as a possibility in Nature. Essentially, I told myself, "Just go with it." Once I accepted this I asked myself what else this could imply if 0=1 is an expression of reality.

There are a host of physical phenomena yet to be described or explained properly at both the quantum and cosmological levels. Mysteries abound on both ends of the size spectrum. My argument is, logical or not, if we plug in the equation, 0=1, then many, if not all, of these phenomena suddenly make sense. This "leap of faith" has been required for many theories in the past—including gravity and relativity—a period of time between proposal and experimental proof in which theorist and his or her handful of believers say, "Just go with it and see." This will be a recurring theme throughout this paper because...

It turns out to imply quite a lot.

Since the initial publication of this paper, I have refined and clarified my thoughts, and deleted many errors. Previously, I would repeatedly give lengthy, elaborate explanations every time I mentioned the concept of 0=1 and, particularly, its resulting field. Moving forward, the idea as a whole I have now labeled O-theory, and the previously named the 0=1 field as the O-field. The symbol O was available in physics due to its similarity to (and the likelihood of being mistaken for) 0. Well, conveniently, regarding O-theory or in discussing the O-field, if the O is mistaken for a 0, no big deal, since the theory is rooted in 0=1. So, O-theory, 0=1-theory, or even just 0-theory, there is little problem with misunderstanding or confusion.

A couple of illustrations, one new, one familiar, help to demonstrate the concept. The number of illustrations is actually limitless, but we will use two for now. If there is a table with a book on it and I ask, "How many books are on the table? "You would answer, "One." If I remove the book and ask you the question again, you would answer, "Zero. None." Then I ask, "Between the moment the book was on the table and the moment it was not on the table, how many books were on the table?" There is no logical answer except for either "both zero and one" or "neither zero nor one." And both answers are correct. For a fraction of a moment, so small as to equal zero duration, there was both one book on the table and no book on the table. Zero and one occur simultaneously in that "in between" moment. The more familiar illustration is Schrodinger's famous cat. Before we open the box, we do not know if the cat is alive or dead. Upon opening the box the cat dies for certain. Even assuming the cat was still living before we open the box, the process of opening the box renders him an ex-cat. It is the opening of the box itself that creates the in-between moment when the cat is both alive and dead. In that infinitesimal "between" the cat is both 1 (live) and 0 (dead). The in-between moment in both instances, the book and the feline, is indivisible by any number. Its duration and dimension are equivalent to $0=1$. On both sides of that moment we have either 0 or 1. At the moment itself, we have both.

Another way to describe $0=1$ is it is the point/moment where all discrete particles, charges, and forces attain a state so small and brief ($0=1$ dimension and $0=1$ duration), as to be no longer discrete. At the $0=1$ level, they "merge" into a field—the O-field. The individual particles, charges, and forces become indistinguishable one from another and form the O-field, akin to but distinct from the otherwise discredited concept of the "ether." Let me repeat that, akin to but distinct from the "ether." The O-field is NOT the ether, as it natural philosophers conceived it centuries ago. The ether makes its way back into popular thought from time to time. Even Einstein toyed with an idea similar to the ether. The O-field, unlike the ether, is not a "substance" through which celestial bodies pass, or on which the cosmic drama is played out in time. The O-field *is* the drama; it *is* the substance; it *is* the celestial bodies; it *is* time. It *is* everything.

Back to $0=1$. Many phenomena can be explained by a reality in which $0=1$. Here, however, are a few, listed in no certain order except that in which they popped into my head or I came across in my reading:

Support for 0=1 can be found in Robert Laughlin's *A Different Universe*, particularly Chapter Nine. "Such a length [one which renders the background electron density to be infinity] conflicts fundamentally with the principle of relativity, which forbids space from having any preferred scales. No solution to this dilemma has ever been found" (104). The value of the ultraviolet cutoff remains unknown. I suggest it can be found in 0=1. He goes on to discuss postulates that are unfalsifiable, and therefore irrelevant experimentally. Laughlin writes, "The accepted practice of declaring unmeasurable things to be nonexistent— even when the problem lies in one's own experimental shortcomings—then makes the issue mood" (111). Yet, that is with current technology. I propose that 0=1 will one day be verifiable. In the mean time, nearly every example given by Professor Laughlin (including the correlated-electron effect (152)) seem, at least to my mind, to be explainable, if 0=1 is a reality. Certainly nothing else explains them, and they are many.

Laughlin also, perhaps ironically, sums up the biggest problem with 0=1: it is (currently) unfalsifiable. He writes: "If renormalizability of the vacuum [of space] is caused by proximity to phase transitions [0=1], then the search for an ultimate theory would be doomed on two counts: it would not predict anything even if you found it, and it could not be falsified" (154). Alas. I hold out hope that technology will catch up one day.

0=1 is equivalent to Richard Dedekind's concept of *a*. Though in the case of 0=1, what is being described is not an abstraction, but a real event, a point/moment that is the foundation of reality itself.

The fact that we have harnessed, manipulated, and utilized to human purposes this 0=1 enigma (unaware that that is what we had done) in our electronic and wireless technologies is at once testament to human ingenuity and the limits of our examination of reality.

Just one example of this technology: 0=1 is the explanation of the creation of "mini Big Bangs" in experimental settings. Huge facilities are required. Massive amounts of energy are required. Yet in comparison to the energy of the Big Bang, which was infinite, these amounts of energy are minuscule. However, it still demonstrates 0=1 in action—something from nothing—as a result of massive energy. 0=1 is a fundamental occurrence. It is happening everywhere all the time on an incomprehensibly smaller scale—a scale at which 0=1.

0=1 explains why there is no such thing as "before" the Big Bang. There is 1 which represents, as one person more clever than me once put it, "Life, the Universe, and Everything." There is 0, which is absolute nothingness. Then there is 0=1, the moment when they were/are the same. There is no such thing as "prior" to 0=1. 0=1 is not the same as just 0. Just 0 is nothing. And

1 is something. In between is 0=1. Shake your head all you want. Because I have many more phenomena to list that, if 0 does not equal 1, we as yet have no explanation for. If, however 0=1, many things suddenly make sense.

0=1 explains the one directional linear nature of time. With both 0 and 1 there is no reversal. There is either stasis, or forward movement in the form of a whole number. There is no negative in the equation, no going backwards, nothing before 0. There is no negative, and nothing precedes 0 except in abstraction. There is no "prior" to 0=1, and the fundamental nature of Nature is to move outward (and inward, but more on this when I discuss the Will to Equilibrium). The same goes for Time, one direction only—forward. Mathematicians will argue, "Of course there are negatives!" and then gladly show you a list of equations in which a minus sign is the star of the show. But those are human constructs. In Nature, negatives do not exist. Physicists will say, "There are positive (+) and negative (-) charges!" Again, those are manmade symbols created to describe attractive and repulsive forces that do exist in Nature. But positive and negative, and their accompanying symbols, are only humankind's way of describing those forces. They do not really exist except in a dictionary or a physics book.

The energy inherent in the equation 0=1 is incomprehensible, though it is very real. Consider, how much energy would it take in an experiment to make the mathematical equation 0=1 a reality? How much energy would be required to make 0 become a 1 or 1 become a 0 simultaneously? The answer can only be an infinite amount of energy. Where and when was this infinite amount of energy a reality? At the Big Bang. The cataclysmic start to our universe—and what other could it be than cataclysmic?—was the explosion, the Big Bang, that resulted from the infinite energy and pressure contained in the singularity, more accurately expressed by the equation…you guessed it: $0=1$.

The result of the Big Bang was our binary universe, as we experience it, in which we have 0's or 1's. At the moment of the singularity, 0 and 1 were the same. Immediately after the event, the energy level was no longer infinite. From 0 had emerged 1 and we have had a binary universe ever since. $0=1$ became 0 and 1 in a four-dimensional universe, the "on" or "off," "is" or "isn't," "Being" or "Nothingness" that we experience every day. More on this when we come to the Will to Equilibrium. You're going to love it!

We experience existence our four dimensions as binary because, in our reality, something either   is or is not, there is no in between, except at the point/moment $0=1$. With $0=1$, this equation expresses how it is both. It is and is not at the same time. It is Schrodinger's cat in a nutshell. It is both alive and dead, but it is only alive or dead upon observation. This is the due to the limitations of our instruments of observation in a macro world. We humans and our tools and methods are limited. Nature is unlimited.

Here I introduce a new term: maxiquantum. Go ahead, get your giggle over with. Better? Let's proceed. What is maxiquantum? It is the theoretical smallest non-zero particle. Infinitesimally smaller than the Planck length, and irreducible, it is the smallest possible entity on the positive side of 0. Reduction of the maxiquantum results in zero with no possible quantity between. Mathematicians are going to LOVE it. I was told over twenty years ago, and reminded again quite recently, that if we reduce to decimals infinitesimally until there is no longer a distinction between a theoretical "smallest non-zero" particle, that, mathematically speaking, there is no distinction between that theoretical "maxiquantum" particle (my term) and actual 0. This is where Nature and mathematics part ways. Yes, mathematically speaking, there is no distinction. Nature, however, say, "Oh, really? Watch this."

More than one person has encouraged me to use the term "infinitesimal" or some new or novel variation or spin on that word. However, infinitesimal indicates ongoing diminishment infinitely approaching, but never reaching actual 0. What I am proposing is that there is a point/moment where we reach $0=1$. It is not true 0. As already stated, *that* would be absolutely nothing. Neither is it infinitesimal. It does finally reach a point of culmination. In conventional math, everything between 1 and 0 is expressed as a decimal. A point, a near-infinite string of 0's, and a 1 somewhere at an imaginary, non-exist "end," never reaching actual 0. Nevertheless, even the hypothetically smallest non-zero decimal is still equivalent to a 1, decimals be damned. If it is measurable, if it has dimension, if it exists, then it can be ascribed the digit 1. My argument is there is a point/moment that is neither 0 nor 1 nor some decimal in between, but, rather, 0=1. Again, not possible in conventional mathematics. Possible for Nature.

More phenomena in support of 0=1: Such a concept supports the existence of non-discrete (subquantum and maxiquantum) particles, like quarks, muons, leptons, and those yet to be proposed or discovered. There is a continuum of quantum particles of smaller and smaller size stretching from electrons down to the 0=1 point. It is self-evident that such particles must exist. Historically, detractors have repeatedly denied their existence. Yet theorists keep proposing them. Researchers keep discovering them. While many say, with each discovery of a new particle, "Well, that's it. We've reached the end. Can't go any smaller," Nature says, "Yes, I can. Way smaller."

0=1 unites Newtonian physics, general relativity, and quantum mechanics. Yes, you read that correctly. The concept of 0=1 in part, and O-theory as a whole, is *that* theory, the theory everyone has been pursuing for over a century. O-theory addresses the infinitely small, the infinitely large, and the ways in which they interact with one another, in both Newtonian (classical) and Einsteinian (relativistic) ways simultaneously (more on this in a later section).

$0=1$ justifies why the universe is so large. Energy equals mass. Einstein proved this. The amount of energy to make $0=1$ a reality in the physical universe would have to be an infinite amount of energy. This is what we see at the Big Bang. An infinite amount of energy would make for a massive universe (as it *has*), with a considerable amount of energy leftover—energy that has not converted to mass and is undetectable except for its influence on the universe in its ever-accelerating expansion. We call it dark energy. For this reason, we have a massive universe comprised of finite mass but infinite energy and no boundary. The infinite energy at the Big Bang is *still* infinite; none has been lost. We just can't see it.

$0=1$ explains the *how* of Einstein's famous equation: $E = mc^2$. It is the equal exchange of energy and mass, both being flip sides of the same coin. Energy is mass; mass is energy. $0=1$ is the symbolic expression of energy becoming mass and mass becoming energy in the "in between," when they are simultaneously both mass and energy.

$0=1$–is apparent in all chemical processes. In every chemical conversion formula, we rely on the mysterious and vague term "yields" and its right-pointing arrow, to explain the process. This, for centuries, has been satisfactory. It is no longer. What is *really* happening during the point/moment of conversion from one state to another, from individual element to molecule, and so forth? This is a 0=1 point/moment at its finest. The point/moment between the water molecule as it goes from liquid to gas. At 0=1, it is *both* liquid and gas, and *neither* liquid nor gas. It is the point/moment between bonding of elements. Is it two hydrogen atoms and one oxygen atom, or is it a water molecule? At 0=1 it is *both* and *neither*. It is the "in between" state. And it takes place in a point/moment that is equal to 0=1. On either side of the event, on either side of the $\rightarrow$, we have only one entity/expression/result of the chemical event. In the "in between" point/moment of 0=1, they are both and neither at the same time. I am reminded of the cartoon in which a physicist has drawn an arrow in the middle of an equation and written, "Magic happens here." His colleague in the cartoon explains to him, "That's not how it works." But for more than two centuries that is exactly how it has worked: that little right-pointing arrow in chemical equations has been our equivalent to "magic." That need be the case no more. There is no magic. There is simply Nature making a 0 equal to a 1.

Consider transition state theory (TST). The idea that in catalysis there are the reactants on one side of the process and products on the others. In between are the substrates which are neither one chemical nor another, but an intermediate state that can continue through the process to become a product, or return to the original chemical make-up and be indistinguishable from the initial reactant. This in-between state is the chemical expression of 0=1. Both reactant and product, yet neither reactant nor product, at the same point/moment (McFadden and Al-Khalili 70-72). This is the essence of 0=1. It requires only the boldness on the part of scientists and philosophers to consider it as a possibility.

Consider, for one more example, the conversion of O_2 and C to CO_2 in animals, and vice versa in plants. There is a point/moment when they are neither and both. If we ascribe a moment of measurable duration, however near-infinitely small, we still have either carbon and oxygen on one side of the moment and carbon dioxide on the other. At the 0=1 moment, the point where Nature works her "magic," we have both and neither, in a length of time equivalent to 0=1. It is immeasurable because it occurs in a measure of time and dimension that practically does not exist. Yet, it is very real. We can see it in almost *every* phenomenon, if only we will look.

0=1 and its resulting field—the O-field—is a fundamental phenomenon that comprises the very fabric of the universe. It is the interface, the threshold, of existence and non-existence. We can, therefore argue the universe is composed of existence and nonexistence *simultaneously*. Those scientists and philosophers who argue for a reality in which things like stars, planets, humans, brains, neurons, and atoms do not *really* exist, due to quantum mechanics, could be right. Those who take the Newtonian route, that such objects empirically *do* exist, could also be right. A universe in which 0=1 is reality proves *both* arguments are correct.

Though simple and elegant, 0=1 is *quite* profound. Yet, formidable resistance to it remains. The phrase 0=1 appears in reputable science journals and books as little more than a joke, at which the authors chortle and guffaw, impressed by their own wit, not realizing what they have put forth in jest may in fact be *the* answer.

Theorists like Alan Guth have taken us back to one trillionth of a trillionth of a trillionth of a second in time to a point one trillionth of a trillionth of a trillionth of a centimeter in size. His theories have recently been proven out by the discovery of gravitational waves. The concept of 0=1 takes what they accept as "true" to its logical conclusion. Many physicists and cosmologists are all comfortable with "near" infinite energy and "near" zero sizes. I am merely saying Nature goes all the way. It is her final trick: *actual* infinity and *actual* zero in a point/moment in space-time…before there was either space or time! The moment where *everything* and *nothing* were one and the same—0=1.

A saying goes that mathematics is the language of the universe. Another says that Nature is mathematics, and mathematics is Nature. The universe and mathematics are the *same*. Mathematics explains *everything*! These contentions are true. Somewhat. I know that by any measure of known reality or sanity, for that matter, that 0=1 is mathematically impossible. However, Nature does not give a damn about mathematics. Reality transcends mathematics. Mathematics proposes measures of length and time that are mind-bendingly small, yet they resist the fact that Nature goes one step further and achieves 0=1. Mathematicians and physicists scribble and sweat and struggle with their μ's and τ's and ε's, and all the while Nature just smiles. Conventional mathematics cannot solve the mystery of 0=1. Only after the Big Bang, less than a Planck time after that 0=1 point/moment, were mathematics and calculus born. The equations necessary to describe the laws of Nature would, in time, grow as astronomical and complex as they are astounding and awe inspiring. At the singularity, the known laws and forces of Nature break down. So do conventional mathematics. Conventional mathematics and calculus, I am sorry to say, are limited. For mathematics is a human construct, much like religion. However, the "god" at play here is far more clever than any human, and her name is Nature. She had some "mathematics" of her own, a hidden ace up her sleeve, one additional equation she tried long and hard to keep secret: 0=1.

It seems self-evident to my mind that a theorem describing all of reality and its origin must itself lie outside of anything it could contain. In other words, the theorem must be contradictory, mathematically impossible, illogical, a paradox. $0=1$ meets that criteria.

I first proposed $0=1$ exactly 20 years ago. I have it written in a note in a journal from 1996. But I quickly rejected the idea. I caved to minds greater than mine and followed the mathematicians who said such a thing was impossible. However, 20 years later, a few weeks into the composition of what is now Section I (the only section I had any concept of writing), when I was in my classroom, conference period, feet on my desk, reading the penultimate scene in *Deadpool and Death Annual* (1998), where Deadpool is between this life and the next, inhabitant of both worlds, dead and alive simultaneously. I sat up in my chair with mouth agape. An "old friend" I had abandoned 20 years earlier came rushing back to my mind: $0=1$! Convinced in 1996 that such a concept was impossible, I had not thought of it since then. "Hold up, wait!" a voice cries, "You're telling me this so-called profound idea came to you from a Deadpool comic book?" "No," I say, "It came BACK to me from a Deadpool comic book."

Upon revisiting $0=1$ in my investigation and musings over consciousness and the behavior of the electron, I am now convinced more than ever that $0=1$ is a fundamental fact of Nature.

The initial reaction to hearing, "0=1", will be laughter, and I may never live to see it proven. However, I believe this is the *only* answer. It explains far too many phenomena for it to be untrue. It is on this point that 0=1 finds the weight of its argument, not in its mathematical or experimental provability. As a man of no faith, it is ironic that what may prove to be my contribution to the ideas of mankind, and a testament to the power of the mind, at this point in time, requires *faith* to believe.

IV: THE WILL TO EQUILIBRIUM IN THE UNIVERSE

What do I mean by the Will to Equilibrium in the universe? Just that. Everywhere in Nature we see the push and pull of opposing forces engendering the grand reality we inhabit.

However, let me be clear about one thing. Though the "will to equilibrium"—we could call it the "inclination" or "urge"—are all words and phrases that imply personification and thought on the part of celestial bodies, if not a personal, sentient, cosmic "being"—a god—this is *not* the case. The Will to Equilibrium is itself a force that is fundamental to the universe. It is difficult to avoid ascribing terms like "will" or "urge" or "inclination" without implying personality. The Will to Equilibrium is impersonal. Its existence and its influence are no more personal or sentient than is a magnet drawing lead shavings to itself. The Will to Equilibrium just *is*. However, its *is*-ness is what underlies all of Nature.

The Will to Equilibrium is the balance sought by Nature in the struggle between symmetry and entropy. Although disorder is easier to achieve through entropy, survival and maintenance of order seems equally to be a natural state of the universe. Spontaneous order itself is an expression and result of the Will to Equilibrium. Balance and the elegant "design" of simple molecules and crystals is symptomatic of this universal "desire" for order in defiance of entropy.

Equilibrium is a natural state of the universe. It is why 0=1, and the resulting O-field, must be a foundation of reality. The universe strives for equilibrium because it is a fundamental element of Nature—symmetry and entropy contending with one another until equilibrium yields a singularity again at which 0=1. They are one and the same. The urge to maintain optimal conditions are fundamental. It is the pull of symmetry versus the push of entropy that creates equilibrium, which creates existence.

If a will to equilibrium (lower-case intended in this instance) exists, then it explains many, if not all, phenomena in the universe. It explains the smallest and largest entities, from the quantum realm to the expanse of the cosmos.

The Will to Equilibrium explains why we inhabit a binary universe. Every phenomenon has its equal and opposite counterpart. On and off. Alive and dead. Dark and light. Positive and negative. In between every opposing pair of phenomena lies 0=1. At the singularity, all of existence collapsed from a previous universe into a point/moment equivalent to 0=1. From that singularity, when 0 and 1 were indiscrete, the universe exploded into 0's and 1's. All of existence is the interplay, the dance, of those 0's and 1's, the yin and yang, pushing one another apart and pulling one another together. The binary universe "wills" to be a singularity again.

The four currently known forces in the universe are themselves indicative of the Will to Equilibrium. The repulsion and attraction of electromagnetism, the strong and weak nuclear forces keeping atoms intact, the positive and negative charges of particles, drawing them together or forcing them apart, the push of energy and the pull of gravity, all find their basis in the Will to Equilibrium. Any forces we discover in the future will exhibit the same quality.

The Will to Equilibrium explains why planets and solar systems form from chaotic balls of gas through a phase of pure destruction and cannibalism of one planet against another to settle finally in an elegant dance of orbits.

The Will to Equilibrium presupposes and, therefore, "confirms" (in quotations, because they have already been experimentally confirmed) the existence, the *necessity*, of antiparticles. Simply put, in a universe where the Will to Equilibrium is real, there *must* be antiparticles. How could there not be? The fact we know they exist does not affect the self-evident nature of their existence. If we knew of a Will to Equilibrium prior to knowing of anti-particles, then their existence would have been a matter of simple conjecture. There are particles. There must be anti-particles.

James V. Stone, in his study of the work of Claude Shannon, *Information Theory: A Tutorial Introduction*, writes, "[T]he efficient code hypothesis suggests that evolution of sense organs, and of the brains that process data from those organs, is primarily driven by the need to minimize the energy expended for each bit of information acquired from the environment" (3). This "good trick" of evolution, seen in all creatures with brains, is also indicative of the "arrows" of evolution (and time) to result in observers/detectors capable of examining themselves, and thus, of the universe looking at itself. This simultaneous expenditure and preservation of energy in beings meant to observe themselves *and* the universe is no less than the Will to Equilibrium at work.

For this reason, I propose that Jeremy England's concept of the dispersal of energy being the "prime directive" of the universe is only half of a theory

(Wolchover). The Will to Equilibrium, the push and pull between entropy and a desired singularity, between repulsive energy and attractive gravity, is a more complete theory. The dispersal of energy and the generation of energy at the same time is, actually, what is happening. The amount of energy in the universe is constant. If Nature is dispersing it in one place, then she must be gathering it in another. Dispersal and generation of energy occurring simultaneously in order to maintain a sum total of energy equivalent to its origin, which was (and remains) infinite—having begun at point of equivalent to $0=1$.

I know I run the risk of real scientists and philosophers levying the charge of promoting metaphysics or, in modern parlance, "woo" science. Support for the Will to Equilibrium, though not called by that name, makes appearances (perhaps inadvertently) in a number of scholarly works appears that manage to remain tethered in empirical science, and not launch off into "woo" or metaphysics. I mention only here two of them.

Charles Seife's *Decoding the Universe* has much to say about the import of information as an entity in the universe. Again, Peter Watson does a masterful job of summing him up. "'Information is as real and concrete as mass, energy, or temperature. You cannot see any of these properties directly, but you accept them as real. Information is just as real.' [He] goes on to highlight how, when Claude Shannon published his seminal paper, in which he realized that information could be quantified and measured, he also realized it was intimately linked to thermodynamics. There is something about information that transcends the medium it is stored in. It is a physical property of objects akin to energy or work or mass…Nature seems to speak in the language of information'" (399). As good as Watson's summation is, Seife's complete work is worth reading in whole.

Support for the idea of a Will to Equilibrium. Another source is the 1274-page opus of Stephen Wolfram, *A New Kind of Science*, self-published in 2002. Once again, Peter Watson: "The idea that a few simple rules can lead both to great complexity and to order down the line, that order and complexity are different sides of the same coin, is for him the most important thing about the universe, because the universe itself is both ordered and random, containing both simple features and complex ones. For Wolfram, these simple rules explain everything" (411–12).

NOTICE: Here I definitely verge on metaphysics. However, I make no apology, because, if I prove correct, then I remain free of that failing. I have previously discussed the Big Bang and its nature, not the events a trillionth of a trillionth of a trillionth *after* the Bang, but the Bang itself. That singularity, the thing that was before *any* thing was, in a time before Time, has eluded explanation. All math and physics break down and theorists throw up their hands in defeat—that phenomenon I have here attempted to elucidate. Now, I have, perhaps imprudently, given that heretofore *mystery* a name.

I call that confluence of mass, energy, temperature, gravity, and information into a unity the Prime Will.

Even in its present form, spanning fourteen-billion light years, comprised of the O-field, the universe itself, it is *still* the Prime Will. All expressions of will—including all the levels of the Spectrum of Consciousness, including my typing of these words at this moment—are lesser versions and derivatives of a greater force.

When I speak of the Prime Will, am I speaking of God? No. Merely —merely!—that at the 0=1 point/moment of the Big Bang, the singularity contained not only all the mass and energy and gravity of the universe in a unity, but that it also contained all of the information in the universe. Every "bit" or "qubit" (to use somewhat misleading terms) of information that would lead to the creation of galaxies and stars and planets and plants and dinosaurs and humans—every possible equation, every permutation, every pathway, every algorithm, and every cause-effect relationship for everything that would eventually occur in space and time was contained in that unity. Omnipresence—space. Omnipotence—energy and gravity. And Omniscience—

information. "How is this not God?" one might ask. Because a god or deity implies personality, identity, and choice. The Prime Will remains impersonal, a force, not a being. Remember the magnet and the lead shavings.

I first coined the term the First Will over 20 years ago, in the same journal and within the same few pages that I first proposed 0=1. The word "First", as the idea hit me again, seemed too much to imply "causation" in the Western sense of time and linearity. I immediately hit upon the more accurate term "prime." And here we are.

Though this section on the Will to Equilibrium is deceptively brief, it may prove to be the most important idea in the paper.

V. OTHER IMPLICATIONS

General Overview

If variable electron reactive behavior (VERB) is the origin of consciousness on Earth, if the 0=1 point/moment comprising the O-field is real, and if the Will to Equilibrium is fundamental to nature, then there are several more intriguing implications.

If VERB is the first level consciousness on Earth, and the subsequent narrative I have suggested is true, then Darwin's theory of evolution through natural selection applies to behaviors and benefits at the atomic and subatomic levels, extending Darwin backward to the quantum level, a billion years or so before the gene was even an actor in the play. The gene is not the primary player in evolution. The gene merely replicates on a far more complex scale what is happening at the quantum level. The molecules formed at the sub oceanic volcanoes on Earth four billion years ago, the ones who would go on to form life, were the winners in an early game of natural selection. Darwin, it seems, was even smarter than anyone thought, including Darwin.

Nature loves spectrums. All of existence occurs on a spectrum, from 0=1, to maxiquantum particles, to electrons, to atoms, and on, to massive galaxies, to the universe itself, which is boundless. 0=1 demands that Nature exist as a spectrum because between 1 and 0 represent the opposite ends of the spectrum of literally everything. Yet Nature achieves 0=1. Only at 0=1 would the energy of a compressed universe been equal to infinity. Anything short of 0, anything on the positive side of 0=1, in the "in between," and the amount of energy at the singularity would have dropped, however incomprehensibly small, below infinite. And then, the universe we observe today would not exist. At absolute 0, nothing exists. But at 0=1, we have everything and nothing at once.

Both Marconi and Tesla were laughed at for their idea of transmitting information through "empty space." The O-field concept is the logical continuation of their thought, only reducing the field with which I am dealing to a size that runs contrary to conventional math and questions the Planck length, Planck time, and even the speed of light as the "limits" of physical reality. I'm no Marconi. I'm no Tesla. I have zero grounding in mathematics and only a fundamental grasp of physics. All I have is imagination and conviction that, as crazy as it may seem, 0=1 is THE fundamental aspect of reality. It is the equation Nature kept hidden from us until a lowly English teacher in East Texas discovered her secret.

Everyone may laugh this idea off because it is not disprovable, it's unverifiable, cannot be mathematically proven. It can never be tested in the lab or experimental research facility because the math is impossible. I say, "Look around." If 0=1, it should explain everything logically. I believe all of Nature is pointing at herself and pointing at the equation, 0=1, and crying out, "Look! It's right there. It's me." Nature herself is the proof of the equation. No need for a particle accelerator. No need for a lab. No need for mathematics.

The Big Bang and the Big Crunch

The reason the universe is expanding at an ever-accelerating rate is because the repulsive force of the infinite energy at the singularity at the Big Bang is still exerting its influence. The Big Bang is still banging. The universe is still exploding and, because it is so spread out, it is meeting less and less resistance as the fields of these subquantum and maxiquantum particles grow thinner and thinner and yet never drop below infinite. Also, the rate of speed of these maxiquantum particles at the edge of the universe is so fast that time for them is essentially at a standstill. Although it has been 13.8 billion years for us since the Big Bang, for these outermost particles at the edge of the universe, little or no time has passed.

The universe is both accelerating and slowing down simultaneously, sort of, if 0=1. It is expanding due to the repulsive force of its energy. It's contracting through the force of gravity at local points throughout the universe, comprising black holes, new stars, and new galaxies. Because the whole of existence is interconnected on the 0=1 level comprising the O-field—absolutely no part of space is separate from the rest of the universe—such events as black holes, the vacuum cleaners of Nature, could account for a Big Crunch. The universe may be expanding at an increasing rate, but Andromeda and the Milky Way are on course for our bigger neighbor to devour our galaxy in about 4 billion years, with their respective massive black holes to merge into a supermassive black hole.

Why does the universe appear to be expanding outward in all directions no matter from what point you view it in space? It would seem to us accustomed to our terrestrial life that if there was an initial point of an explosion, it should appear that the universe is spreading outwards from that one central point, moving away from us in only one direction, or possibly fanned out like a shotgun blast, but still with definite momentum in one direction, not every direction. This is not the case for a couple of reasons. First, because it is space itself that is expanding, not just the stuff contained in it. Second, as the universe expands, we are traveling right along with it. Because of the infinite amount of energy at the singularity, the 0=1 point/event, the Bang is still banging, and we are riding along with it, relative to everything else. So, no matter what point in the universe we look out from, the energy and matter is still spreading out, away from us, because of the nature of the fabric of the universe, down to the 0=1 level, is going outward. Only at points where gravity has gotten an upper hand at locales in space and formed galaxies and stars and planets do we see the fabric of space warped by the particles held in check by the gravity of the large celestial bodies.

Because the universe began at a point where 0=1, the nature in which a fabric created from infinite energy spreading outward from a point that was once 0=1 would mean that essentially every point in space, as it expands, is also the point of origin of the Big Bang because all of space and time are connected like a fabric at the 0=1, or O-field, level. So, essentially those old cosmologies that placed Earth the center at the universe were not wholly wrong, in a way of speaking. Every point in space is the center of the universe if 0=1.

This is strange, I know. At the singularity, a point where all of existence and nonexistence met in a point/moment of both infinite energy and 0=1, the tension present caused it to explode. Every particle, every single one, formed by the explosion and its aftermath, essentially remained, from its own perspective, "located" or "positioned" in its original point in space, in reference to everything else. The fabric of the universe is composed of 0=1 points/moments that are mimicking and reenacting the Big Bang on a microcosmic level throughout every fraction of space. Therefore, no matter where you are in the universe, everything else appears to be racing away from you. The Big Bang is still banging. If we could teleport ourselves 13.8 billion light years in an instant, stop, and look around, the Cosmic Microwave Background would still be 13.8 billion light years away. How beautifully strange. Consequently, each and every one of us, literally, is the center of his or her own universe, and reaching the "edge" is an impossibility.

Because of the sheer force of the energy at the Big Bang, regarding the will to equilibrium, the struggle between symmetry and entropy, entropy would more logically be the norm in Nature, making gravity and creation a mystery. But that mystery is explained in the Nature of the universe wanting to be a singularity. Again it is a fundamental aspect of Nature for the 1 and the 0 to become a 0=1 again. That is the 1 of all existence simultaneously becoming equal to the 0 of nothingness. Therefore, entropy appears to be more common in Nature than symmetry, though both exist and are pulling and pushing against one another, creating and destroying entire worlds in a cosmic contest of forces.

It is possible in a Big Crunch that we could return to a singularity of infinite energy. Although matter is finite, the original energy of the universe remains infinite, though vastly dispersed in the form, mostly, of dark energy. Only about 4% of the universe consists of detectable matter. The rest of the existing universe consists of dark energy and dark matter, which is potentially still infinite energy. As the universe compresses under ever-increasing gravity, energy would convert to matter that would eventually again be infinite once it was a singularity—0 and 1 having again become 0=1.

In the event of a Big Crunch, the final moments would be the exact reverse of the Big Bang. Expansion of the universe that occurred faster than the speed of light would be reversed in a final-moment collapse also faster than the speed of light. Until that moment, the universe keeps expanding. It is like a giant balloon that keeps expanding and expanding yet never breaks. There is no such thing as empty space in the universe. Even the farthest reaches of the universe are still connected by maxiquantum particles and a "fabric" of 0=1 point/moments that comprise all of existence, particles far smaller, comprising a field far more vast and interconnected than even Professor Higgs predicted. This is the O-field. Even as large as the universe gets, no point in the universe is disconnected from any other point. Therefore it is possible, as massive galaxies continue to form with massive black holes at their center, that the universe, even as large as it is, under the combined force of those galaxies' gravity could slow its expansion, stall, and reverse, drawing the universe back inward, at some point which it would collapse at a speed faster than the speed of light—like a rubber band pulled to its tightest point, then snapped back—resulting in another singularity, a point/event where all four forces converge and essentially cease to exert their influence as the universe converts to infinite mass, thus effecting another Big Bang.

Finite mass becomes infinite mass when all remaining dark energy is converted and compressed under the force of gravity, at which point the elemental forces of Nature cease to exist. The universe would again be nothing but pure, infinite matter and energy at a point of 0=1 dimension and 0=1 duration. And when 1 and 0 become 0=1 again, the process starts over with another Big Bang from which infinite mass converts to energy and erupts into the elementary forces that we see exhibited in Nature and a new universe is born.

Gravity

Now to consider gravity. A possible reason why gravity is the weakest force and yet operates on the grandest scales is because the infinite energy present at the singularity is a far more destructive force than gravity. The tendency of the universe to revert back to a singularity, in keeping with the will to equilibrium, is very strong, but almost insignificant to the tendency of the universe to explode from the point of the singularity where energy was infinite. The universe and its desire for symmetry and to return to a singularity is still all but insignificant in comparison to the force of the energy contained in the singularity and released at the Big Bang.

It is only at the 0=1 point/moment of the Big Bang–in that moment of 0=1 duration and 0=1 dimension when all of existence had its beginning–that gravity ceases to exist. For all of existence and time after the Big Bang, gravity exerts its influence. Though it is the weakest of the four known fundamental forces, gravity is the most influential. The instant the 0=1 point/moment exploded, so also did gravity become evident. Gravity is not attraction of the masses of bodies. Gravity is the adhesion of existence. There is no distance or separation between any two points in the universe. At the 0=1 level, everything is still connected and striving to become a singularity again. Energy, most of it dark, causes the expansion, yet can never cause actual separation at any point in the universe. It is all one. It is the O-field.

Gravity and energy vie with one another to determine the shape of the universe and its content. The Will to Equilibrium assures gravity and energy are at play in every point in the universe, vying with one another, like opposite sides of the same coin.

Expansion is the "push" of energy from the Big Bang that, being infinite, continues to force objects in the universe away from one another. Gravity is the "pull" from the Will to Equilibrium, given that the O-field remains unbroken. Everything in the universe is still just as connected as it was at the singularity. We experience the universe in four dimensions as distance and separateness. But that is just an illusion. We observe the warping of space-time more around large celestial bodies because gravity, due to the sheer size of the body, is "winning" the tug-of-war between the push and pull of energy versus gravity when near a massive body like a planet or a black hole.

Gravity is the inclination of nature to convert back to its original state of infinite energy. Gravity is an attractive force whose goal is to compact, compress, and create energy in the form of heat through friction. However, due to gravity and the mass of the particles which comprise our universe, rather than pure energy, we have planets and stars and galaxies and audacious authors of quasi-scientific papers.

Gravity is the fundamental force of nature that is striving for symmetry against entropy. 0=1 does not express symmetry itself, only the need for symmetry in a binary universe comprised of 0's and 1's. 0=1 does not express symmetry because, at 0=1, symmetry is no longer a necessity. The universe wants to be pure energy again. Only as a singularity is Nature "happy."

Gravity is the "holdover" from the singularity. What was once infinity and zero, everything and nothing, existing in a point/moment before time and space existed (0=1), after the Big Bang began to pull itself back together. Gravity is the phantom risidual of the singularity. It is what keeps the universe "together" in its nascent form, despite appearances in the macro world. Technically speaking, from the standpoint of gravity, the universe, however vast, is STILL a singularly, and I have named it the O-field. Weird, I know.

Gravity is simply a residual result carried over from the singularity at the Big Bang. At the 0=1 level, the universe is still a singularity. In reality there is no such thing as nothing, except as an idea. Gravity is the tension from the Will to Equilibrium's "pull" versus energy's "push". Like a rubber band or balloon stretcher to infinity but never breaking. The warping of space-time around massive bodies increases in proportion to the mass of the body and the distance in space from it, as well as that of other massive bodies working from other locations. The easing is greater near massive bodies because the pull of gravity vies against the push of energy. In localized areas gravity results in stars and planets. But the majority of the universe remains dark matter and (much more so) dark energy. Essentially, gravity is the pull from the singularity that has never actually been broken at the 0=1 level. It's all one big whole. It is the O-field.

We know time runs slower as gravity increases. This is why at the singularity, time almost stops as we approach infinite gravity and infinite energy. At the point of infinite energy, both gravity AND time cease to exist. There is only pure, infinite energy. Yet it is fantastically unstable. Its inevitable result is a Big Bang. The duration of this infinite energy is equal to 0 because there is no such thing as time, yet it's energy is infinite because it comprises what was once the entire universe. This is the essence of 0=1.

Gravity exists at the quantum level. There is even an entire field dedicated to its study—-quantum gravity. Yet what it is in reference to O-theory has yet to dawn upon the leading thinkers. Gravity in general—quantum gravity, specifically—is the pull of the O-field, the seamless unity existing at the 0=1 level, that remains unchanged from the singularity at the Big Bang. Everything is connected at the 0=1 level. It becomes apparent (measurable) only at macro levels, and therefore appears to be distinct to large, non-quantum entities. This is untrue. Gravity, electromagnetism, and the strong and weak nuclear forces are all expressions of a singular phenomenon at the 0=1 level. They are merely four expressions of the O-field remaining intact, despite the appearance of separateness or distance. Only in localities where particles have come together to form macro entities do we begin to identify the force we call gravity. But from the smallest to the largest, from bosons to black holes and back again, the attractions between them (call it one of the nuclear forces, electromagnetism, or gravity) are all just the O-field in action.

Dark Matter and Dark Energy

The sum total of all matter and energy in the universe is equivalent to 1. Yet because 1 is equal to 0 at the 0=1 point/moment it is also infinite. The reason it has remained infinite is because the energy that created it was infinite. However, only a small fraction of that energy over the last 14 billion years has converted to matter with a mass large enough to be observable and gather together to form the celestial bodies we see, as well as the elementary particles we can detect. The vast majority of matter remains dark because it is too small to detect, yet the total amount of matter is still less than the amount of dark energy because the conversion of energy to matter requires gravity, a very weak force in comparison to the others. Dark energy is dark could be because it is so much smaller than light particles themselves, that it is not observable. We can only see its influence. Only light particles emit light. Energy particles smaller than light would have to be invisible—dark energy. However, because the energy at the singularity was infinite, there is a whole lot of it, much more so than the matter that makes up what we call the physical universe, what we experience as space and time, the O-field itself.

Most dark matter is matter simply too small to be detected–exotic, yet-to-be-discovered particles extending along the spectrum approaching 0 until nature reaches absolute 0. I would not be surprised to find there remain a near-infinite number of particles yet to be named and discovered.

Dark energy is the remaining energy from the infinite energy at the at the 0=1 moment of the birth of the universe–the Big Bang. What has not converted to matter remains energy. But it's dispersal across the universe causes it to be invisible. Only energy in bundles large enough to generate heat, and thereby, light, are visible. The vast majority of energy is dark. We think of energy as exhibiting light. But pure energy has no light of itself and is "invisible" when dispersed across all of space. It comprises the field that is the canvas of existence—the O-field.

Now for the kicker: At the 0=1 level, dark matter and dark energy are the same thing. Ask Professor Einstein. Being both 0 and 1, they both 1). are undetectable, and yet 2). exert influence on the surrounding space and the galaxies in that space. Being both energy and matter at once, they exert the repulsive force that is accelerating the expansion of the universe, as well as the gravity that holds galaxies together and keeps them from attaining exit velocities from the larger galaxies they orbit.

To summarize one more time, at the risk of redundancy and sounding suppliant: The amount of energy in the universe is infinite. Most of that energy, left over from the Big Bang, is dark energy. Pure energy emits no light, comprises $0=1$ dimension, and exists for $0=1$ duration. It comprises the whole fabric of the universe—the O-field. It emits no light, yet it does exert force. This is why we see the expansion of the universe. In a billion local events, however, where energy has converted to matter, gravity takes hold, and we have things like galaxies and planets and human beings. We find support for this in the recent discovery that quark fusion produces eight times more energy than atomic fusion. Every time researchers discover and collide smaller and smaller particles, the results are higher and higher amounts of energy and new particles. Yet the resulting new particles also live for shorter and shorter periods of time. Logically at the $0=1$ point/moment, we have pure energy that comprises $0=1$ dimension and $0=1$ duration. Clearly this idea circumvents mathematics and relies on pure logic. And that logic dictates that this is the only answer. It seems self-evident. Science and conventional mathematics will most likely reject this proposal. But there can be no other answer if current knowledge and understanding of these phenomena are taken to a logical conclusion. Nature does not do math. Nature transcends math. I feel I have accurately described what nature is actually doing. We need engineers and designers now to figure out how to duplicate it, harness it, and utilize it to human purposes.

Another reason dark matter and dark energy remain elusive is because particles at or near 0=1 are so small as to almost avoid decoherence entirely. Almost. Like the Higgs boson, hidden for 50 years after its proposal, even smaller particles, though forming the field of reality itself, almost never bump into one another. They remain, almost, in a permanent steady state. Almost.

What scientists term "dark energy" and "dark matter" are collectively the O-field itself. Invisible, yet exerting both repulsive and attractive force—expanding the universe at an exponential rate while also gluing it together. That is the O-field.

Quantum Strangeness

Pure, infinite energy occupies 0=1 space and emits no light. We tend to think of energy as always emitting some form of light energy or its analogous equivalent. But pure, infinite energy does not. And in 0=1 space, with no physical bodies by which to measure distance and motion–i.e. zero relativity–time also does not exist. Therefore its duration at the point of infinity is equal to 0=1. Infinite energy is also unstable. It cannot endure for any length of measurable time because time, essentially, no longer exists. So, energy reaches infinity at the exact moment it reverts back to less than infinity, with 0=1 time elapse "between," resulting in no less than another cataclysmic event…another Big Bang.

All particles in the quantum world are capable of both repulsive (energy) and attractive (gravity) forces. Because of the infinite level of energy at the singularity at the Big Bang, most particles, speeding outward from the "center" of the universe, or zipping back and forth along the O-field, exhibit a repulsive force. In electromagnetism, they exert both forces simultaneously. But where matter has gathered and cooled in significant quantities, gravity, the attractive, albeit weakest, force takes the upper hand. As celestial bodies grow larger— planets, stars, black holes, galaxies—their gravitational fields increase and expand, exerting ever more influence on the shape of space-time surrounding them. But whether at the quantum level or the macro level, we are really witnessing the same behavior in the smallest as well as the largest entities. The difficulty in reconciling the four forces lies in the inability to recognize that they are merely variations of one fundamental force: the Will to Equilibrium—the push and pull of symmetry versus entropy, the force that enables existence and non-existence at the same point/moment, all of creation crushed to a point of both infinite energy and absolute nothingness. $0=1$.

Why the wave function collapses into a distinct position or state upon observation: Simply because the instruments used to make the measurement are conceived, designed, built, exist in and are dependent upon a four dimensional universe, as are the non-mechanical instruments that observe the measurements—human eyes, ears, and brains.

Quantum entanglement and O-theory: In our four-dimensional world, we experience time and observe distance. Billions of years' time and billions of light years' distance. But at the O-field level, when considering the field it comprises, neither time nor distance exist. Therefore every point in space is the same, and time is irrelevant. We observe quantum entanglement, two particles light years distant from each other in the four-dimensional world, acting simultaneously on one another. Because at the O-field level they occupy the same space at the same time. Essentially, they are the same particle. At the O-field level, the two particles are one, and thus, by extension, all particles are one. Support for this can be found in the work of Nicolas Gisin (Seife 223, also ff. same page).

The uncertainty required by quantum mechanics becomes even more necessary at the O-field level. When all and none, everywhere and nowhere, is and is not, existence and nonexistence, are one, then uncertainty, and its sibling, potentiality, become indistinguishable. Every "where" is everywhere. Although every "when" is likewise "everywhen" because of the unidirectional linear nature of time.

Indeterminacy, uncertainty, and probability do not rule out causality. At the $0=1$ level, the preceding phenomena are partnered with potentiality. Though the results maybe infinite, they are still the products of causality. Therefore both uncertainty and certainty occur simultaneously. Determinacy can actually be extrapolated from indeterminacy. It is for this reason I feel the need to replace the term "probability" with the more accurate term "potentiality" in reference to quantum phenomena.

Instability increases as we descend down the particle spectrum toward $0=1$. At the quantum level: Instability = uncertainty = indeterminacy = probability = potentiality. They are all essentially the same phenomenon. The essence of the O-field is potentiality. All forms of energy and mass are contained in the potentiality offered by the O-field.

Erwin Schrodinger wrote, "Entanglement of predictions arises from the fact that the two bodies at some earlier time formed in a true sense one system, that is were interacting, and have left behind traces on each other. If two separated bodies enter a situation in which they influence each other, and separate again, then there occurs what I have just called entanglement of our knowledge of the two bodies." He was SO CLOSE! The "one system" is what I term the O-field. And the "earlier time" is the Big Bang. As time and distance are irrelevant in regards to the O-field, no time has elapsed since the Big Bang, and no distance exists between particles since all particles are one. From Newton, to Einstein, to Borh, there is no distinction except in reference to which "world" one is describing—the micro or the macro. They were ALL correct. Between them stretches a seamless continuum that they simply were unable to conceive.

The O-field helps explain quantum entanglement. The particles are not really separated by what we know of as "distance." Support for this idea is found in the work of Nikolas Gisin and colleagues in Geneva in 2000 in experiments involving the "superposition" entangled photons. They found, indeed, that a particle can be into quantum states at once; in addition, a particle can be both a 0 and a 1 at the same time, having simultaneously both up spin and down spin. They also determined that information sent between the entangled particles, if such is the process, would have to travel 10 million times the speed of light (Watson 420). The concept of the O-field explains this because at the 0=1 level, the universe is still a singularity. Information can be shared across what we experience as vast reaches of space because in reality those reaches do not exist.

As we trace the spectrum of entities downward in size from galaxies, stars, planets, cars, humans, cats, plants, cells, molecules, atoms, electrons, to quarks and bosons and smaller undiscovered particles, we gain in number of entities and particles. Far more stars than there are galaxies. Far more planets than there are stars. Far more electrons than there are planets. And far more particles on the end nearer the 0=1 point/moment than there are electrons. As particles grow smaller, their potential energy also increases exponentially from entities further up the scale in size. At the 0=1 scale are infinite particles, comprising the O-field, with relative energies equal to the Big Bang, because they ARE the Big Bang, merely spread across 14 billion light years.

O-theory accounts for what quantum physicists call "preprogramming" of entangled particles. Quantum theory does not account for such preprogramming. O-theory does, because the O-field, despite being spread across 14 billion light years and growing (for now), is still a single, unbroken field. Be it 14 billion light years across, or a singularity trillionths of trillionths of trillionths of a centimeter, the O-field is the same. No point is distinguishable from another at the field level because there is no distinction. Everything in the field IS the field, albeit emerging as macro entities experienced in four dimensions (3 of space, 1 of time).

Electrons appear to be the "interface" particles between the micro (quantum) and macro (classical) worlds. The electron is the "bridge" between the two worlds, acting and interacting comfortably with both the atoms to which they belong and the quarks and smaller particles of which they themselves are comprised. Existing at they do in both worlds indicates that both worlds are really not separate. Only the electron, as far as we know, inhabits both, making it the "glue" of quantum and classical physics.

Yesterday, while discussing O-theory with some students, I made the contention that O-theory is capable of explaining quantum entanglement, what Einstein called "spooky action at a distance."

It does.

One student asked, "Can it explain how a particle changes its state when it's observed?"

I said, "Yes."

He asked, "How?"

And I did not have an answer. I told him, "Let me think about it."

I thought about it.

Here's what I came up with. Could be bull hockey. Time and smarter people than me will decide.

WHY DOES OBSERVATION AFFECT THE STATE OF QUANTUM PARTICLES?

It is simply the nature of existence and our experience of it at the macro level. All of the universe is in a continuous "hum" of quantum uncertainty at the O-field level. Everything is in motion and indeterminate. Only upon observation does an entity become a "thing" that can be measured (either its motion or its location).

At the macro level of our three-dimensional, human world, we see this in something as simple as someone waving their hand back and forth in front of their face. If we focus on their face (or on ANY point other than their hand moving), then the face is in focus while the hand to us appears as a blur, its own little quantum "cloud" of indeterminacy. If, however, we follow the moving hand with our eyes, the hand becomes a "thing" now, clear, determinate, and observable. But, at the same time, the face (or any other point) becomes indeterminate. It is no longer our point of focus or observation. Only what we are observing becomes "real" at the moment.

Which begs the question: Does anything have objective reality when not being observed? The honest answer is "No." When not being observed, things lose their "thingness" and become what they really are at all times regardless of observation: fuzzy quantum blobs of indeterminacy. The key is in realizing who or what is an observer. Anything can be an "observer" of anything else. Even at the quantum level, particles can be "observers" of one another. This is the secret of the photoelectric effect. It is also, if my ideas are correct, the basis of consciousness at the electron level. When an electron reacts to a stimulus, the electron is the "observer" and the stimulus is the "thing" which the electron is "observing" in that moment of interaction.

This appears to be a feature of reality. All of Nature, all the time, is a humming blob of quantum indeterminacy occurring at the O-field level. The O-field, as I have said before, IS reality. It IS the universe. It IS Nature. Only upon observation does it become a "thing" with mass and dimension existing in a time we call "now." And, building upon Einstein's ideas, what is "real" and what is "now" is determined by the observer, thus making reality relative to who or what is looking at it.

So how is a "thing" (say…your grandmother) stable from one observation to the next? If we look at our grandmother, she becomes a thing we can talk to and hug and get a cookie from. If we look away for a moment, she becomes an indeterminate quantum blob. When we look back, she is still our grandmother holding out another cookie. Why did she not change her state?

There are a couple of reasons.

First, we cannot forget that grandma is also an observer. Reality has a stability to it because we are all observers. Not just we humans, but our four-legged companions are also observers. The plants that surround us and provide our oxygen are also observers. Our cells, our molecules, our atoms, and our electrons are ALL observers of the "things" they interact with.

Our macro experience of reality is saved from scattering into a grand state of indeterminacy because everywhere, at all times, there is always some "thing" observing some other "thing" and keeping reality intact.

Second, and perhaps more important, is gravity. Things, like Grandma, maintain their thingness in the macro world we are familiar with because gravity keeps all of the constituent parts from flying apart. Gravity and observation are what create "reality" from the O-field.

How gravity itself is a product of the O-field I have already discussed in another place. Gravity is not a thing separate from the O-field. It is the result of the seamlessness of the O-field. Gravity IS the O-field keeping itself intact, even as our universe (14 billion light years wide) continues to expand. The universe may be growing (for now), but the O-field remains unchanged.

One of the strange things physicists point out about quantum entanglement is that it is discriminating. Particles are "partnered" with one another and exhibit entanglement only with their partners, not other particles in the vicinity. This is misleading and a slight misunderstanding of what is happening, if and only if my concept of the O-field is correct. According to O-theory, "partnering" is a result of observation, not a "real" occurrence. All particles in the universe are non-discrete at the O-field level. Only observation singles out two for measurement. And, of course, those two particles are entangled and exhibit the "spooky action at a distance" that so troubled Einstein.

Again, Jim Holt had this to say in 2018: "However it works, nonlocality has subversive implications for our understanding of space. Its discovery suggests that we might live in a "holistic" universe, one in which things that seem to be far apart may, at a deeper level of reality, not be truly separate at all" (240).

Mr. Holt, sir. Allow me to introduce you to O-theory.

Field Theory

0=1 has intriguing implications for field theory, in particular magnetism. The reason fields exist is because the very fabric of the universe has no space in it. Down to the 0=1 level, all of existence is connected. For this reason, even the Higgs field is composed of relatively large particles. The difference in effects of fields is the amount of mass in different bodies and celestial events as a result of gravity, which is the tendency of the universe to become a singularity again. Magnetism of numerous items, for example, varies from not attractive at all to almost inseparable. The reverse of this, the repulsion power of magnets, is the result of entropy, the pressure and inclination for a 0 to become a 1. Magnetic repulsion and attraction, taken as one, are the very expression, held in our own hands, of entropy and symmetry being elements of the same phenomenon. The makeup of the entities involved determines the influence of the field on a body as it passes through or near the field. Most is undetectable. The fact that magnetism and electricity are one and the same is just another example of a 0=1 event. Magnetism and energy, repulsion and attraction, entropy and symmetry, are all the same. Hopefully by now the picture is becoming as clear for you as it is for me. We know these things to be fundamentals of reality. 0=1 explains how.

Also, regarding field theory, the largest field of all—the O-field—is the one that is undetectable because it occurs at the 0=1 level. It is the very canvas on which all matter and energy exists. The whole universe is a field at its maxiquantum level. It is interconnected throughout all of existence. It penetrates every single thing in the universe because at the 0=1 level it is the universe and it wants to be a singularity again. Whether entropy or gravity will eventually win out is still up for debate. With the field being interconnected, like a fabric comprising the entire universe, again the canvas on which existence is painted, the ability of everything to expand forever is very real. Because of the power and force of the blast at the Big Bang versus the power of gravity to bring it all back together, with all of existence being interconnected, either possibility, Big Crunch or eternal expansion, remains. Again, the universe itself is one, single, unbroken, uninterrupted field with no such thing as "empty" space existing anywhere. The individual fields we experience in our three dimensions–gravity, magnetism, etc.–are merely expressions in particular localities that are reflective of the universe as a whole.

Light itself comprises a field. Infinite observers equidistant in space from a source of light, comprising a sphere of radius x, would all observe the same source of light at the same brightness. Thus light itself is expressed as a field, though visible to individual observers as merely single points.

The O-field is THE penultimate field. It accounts for gravity, magnetism, electricity, and quantum strangeness.

The O-field determines and affects not only space (dimension), but also time (duration), because the O-field

IS spacetime. This explains why the entropy of a massive black hole, like the one at the center of our galaxy, is able to affect the entropy of Earth. The O-field maintains influence throughout the universe. Distance and time are irrelevant at the O-field level.

What scientists term "dark energy" and "dark matter" are collectively the O-field itself. Invisible, yet exerting both repulsive and attractive force—expanding the universe at an exponential rate while also gluing it together. That is the O-field.

My description of the O-field is relative to the macro world, the universe we experience and observe. To speak honestly, the 0=1 corridor is merely a 0=1 point/moment at which time and space (think "duration" and "distance") do not exist. Technically, particles "travelling" along the 0=1 "corridor" (as observed in the macro world) are actually in "no place" and "every place" at once. They are at ground zero of the Big Bang, the "center" of the universe, until observed by pesky scientists trying to crack the mystery of their behavior.

The O-field, alternatively and previously known as the ether, does not actually move, because it is omnipresent. It does, however, expand and propagate due the amount of energy in the universe being infinite. The idea of bodies having instantly an effect on an object due to gravity, however far apart, is misleading. Because it is not actually instant; rather it is constant. At the 0=1 level, the effect of gravity is instantaneous because there is essentially zero separation between the entities acting on one another. All points are one, indistinguishable from each other, at the O-field level. In the macro world, which is anything on the plus side of 0=1, we observe duration and measurable phenomena subject to known laws of physics, both quantum and classical, depending on the size of the bodies involved.

Wave functions are essentially an aspect of field theory. Considering the O-field, what we observe as particles in four-dimensional space are actually all one particle at the 0=1 level. With our current technology, observation of the field itself is not possible. We can only observe its effects.

A spectrum of particles decreasing in size infinitesimally, down to actual 0=1, constituting at that point the O-field, allows for both a unified field theory and particle theory. Any quantity greater than 0=1, however small, makes a seamless field impossible. 0=1 *must* be an aspect of reality and not merely a human construct.

One may find support for my concept of the O-field, among others, in the works of Bohm and Nagel. Specifically, see Laughlin (120-121).

If my proposal of a point/moment of 0=1 is correct, then at a level far smaller even than the Planck length and smaller than the Higg's boson and even more fundamental than the Higg's, another field exists—the ultimate field. I describe it multiple times as the "canvas" on which reality itself is "painted." Its seamlessness accounts for things like gravity, gravitational waves, quantum entanglement, and hand full of other mysteries that still make no sense using conventional methods of description. However, the O-field is SO small as to be undetectable. The O-field is, essentially, EVERYTHING. Being seamless and creating the "bond" between everything in the universe, it IS the universe. At the $0=1$ level, what we perceive as things in 3-dimensional space—stars, planets, galaxies, black holes, humans!—are actually NOT separate at all. There is NO empty space. And, being still connected as one whole in the form of the O-field, we are, in a way of viewing it, STILL a singularity. There is no distinction, as regards the O-field, between the singularity 14 odd billion years ago and particles in four-dimensional space-time separated by billions of light years. To the O-field, they are indistinguishable. I don't want to say the universe and all in it is an "illusion." Nor do I accept any "simulation" hypotheses. But ultimate "reality" from the viewpoint of the universe is far from what we humans experience.

Regarding quantum gravity theory: Once again, O-theory is compatible with quantum gravity, or quantum loop theory, in that it accounts for all aspects of the theory as it stands, merely taking the granularity requirement in quantum mechanics to its logical conclusion, the 0=1 point/moment where particles become non-discrete. The O-field is non-discrete. This is why it accounts for gravity. The O-field and gravity are indistinguishable. O-theory accounts for a spectrum of particles stemming from the non-discrete 0=1 field to discrete but quantum phenomena, on to electrons, atoms, molecules, and the entire macro world. Mathematics and quantum mechanics reject this idea. They insist on the granularity requirement of their models. But a non-discrete field at the 0=1 level completes the theory, fills all the holes, clears up all remaining mysteries, and just makes sense. At least it does to me. This seemingly illogical little concept of a point/moment of 0=1 is all that remains to unify quantum mechanics and general relativity. How long before it is experimentally verified and accepted, or experimentally and finally disproven and rejected is anyone's guess.

In a nutshell: Gravity is the tension, the pull (negative energy), of a seamless field (the O-field) comprised of points/moments in spacetime so small as to be expressed no other way than by the equation 0=1.Expansion is the repulsion, the push (positive energy), of that same field. The simultaneous push/pull of the O-field is exemplary of the Will to Equilibrium that is the reason for existence…the "Why" of reality itself. I can express no more simply than that

Information Theory

This section comprises the latest major update (written in late November 2018) . I apologize in advance for its shoddy organization. It is, at this point, mostly notes transposed from the margins of some of my reading into this document, unedited. It is a snapshot of my thinking as I read. Later revisions and refinements will "clean it up." For now, I just wanted to get the words posted to the paper. Among them, are some of the deepest and most profound ideas I've stumbled across so far. Everything in this section is in reference to Information Theory, in particular the work of Claude Shannon and his predecessors. All words in quotation marks are from INFORMATION: A HISTORY, A THEORY, A FLOOD by James Gleick.

Infinite, pure energy comprising $0=1$ dimension and $0=1$ duration with zero gravity is fundamentally and irrepressibly unstable. Precedent to this, as already mentioned, is infinite Information. Energy, by nature, indicates measurability. Energy occurs in time and space. Information requires no time and no space. Pure information is capable of comprising a point/moment of $0=1$. Information has not source. It has no beginning in any of the forces in the universe. It is itself the source of those forces. The universe begins, exists, ends, and begins again from Information.

One concept generally unconsidered is the potential information yet to come. With exceptions, the general concept is (and sadly always has been for most) our current knowledge either is or is close to complete. Naysayers have cried this for millennia. Yet, original thinkers and discoverers keep proving them wrong.

"Itness" as a component of information exists prior to and even absent of qualities or names. There is a reality, an "itness" to all entities and phenomenon regardless of observation or naming. And it is vast. This is why the vanity of comprehensive knowledge is so adorable in its persistence.

Information and its products (matter, for one) as a form of machinery? The machine makes, itself, those phenomena and products by which the machine is made. Another attempt at defining God?

This is intricately tied to the concept of the O-field being the base/source of Information (the ingredients of creation) because it is Information, reduced to its irreducible point.

The binary nature of the universe is a mask.

The Word exists as a separate thing BEFORE the Thought.

Shannon found Nature's mask. If correct, I lifted the mask.

Words about words and language about language that generate impossible paradoxes are STILL information and exist in reality at some level. Human languages have words that are not directly translatable, yet they exist.

"Gödel showed that a consistent, formal system must be incomplete; no complete and consistent system can exist." Unless that system "dissolves" all barriers between systems. Even the rules of logic become irrelevant because logic is a system that itself gets consumed/assimilated by the superseding system. In a world where $0=1$, all divisions, all choices, disappear because nothing is actually separate from anything else. The idea of separateness goes away. There can be no integers or symbols designed to express distinction or otherness. Even coding as simple as 0's and 1's vanishes. Leibniz envisioned a universal language in which numbers could encode all of reasoning. They could represent any form of knowledge. What he failed to realize was that such a system maintains separateness, which is not an aspect of reality at its most fundamental level.

"Robert Brown showed that random thermal agitation would affect free electrons in any electrical conductor—making noise." What is noise in reference to $0=1$?

"Noise" is the ALL of Information precedent to choice, which is the first step toward meaning. "Noise" and the "Prime Will" are one. Every "thing" that is distinguishable in existence is a result of "choice" which is only possible in a binary universe. But the underlying reality of the ALL is a non-binary unity.

James V. Stone writes in his tutorial introduction to Information Theory, "One particularly intriguing consequence of the final result [of a Gaussian distribution] is that in order for the signal to carry as much information as possible, it should be indistinguishable from pure noise" (127).

"In reference to probability in systems like language, some words have a higher probability of following others than some, and many have virtually zero. A message, as Shannon saw, can behave like a dynamical system whose future course is conditioned by its past history." The same goes for processes like we see in the spectrum of consciousness and the emergence of life. Probability plays a vital role in the process as well as in a message/language. Some levels of order/probability exist for ALL processes. Bet!

"Paradoxical though it sounded, random messages carry more information." And here we have the Will to Equilibrium hiding out. 10-14-2018 7:41 pm. Translation: It's the push and pull of equilibrium that leads to randomness having more information, while meaning comes only with choice and distinction. It would seem logical to be the opposite. But thus is Nature.

Information without meaning comprises the vast majority. What it has is *potential* meaning.

VERB is measurable. Difficult as hell currently, but all stimulus/response/choice/behavior of electrons and the resulting effects up the complexity ladder will one day be readable and measurable and manipulable. Once technology catches up cures will be achieved through "traditional" S-R treatments pinpointing the electrons themselves and seeing the desired VERB result.

"When the pressure and temperature of the system have become uniform the entropy is exhausted." This is related to information prior to its detection as meaning.

"William Thompson said energy is not lost, but it dissipates." But it does not disappear. "Confusion and disorder are entropy's essential quality." Information prior to meaning. Meaning does not occur without a detector, but information is constant and available (potential).

"Entropy thus became a physical equivalent of probability: the entropy of a given macrostate is the logarithm of the number of its possible microstates. The second law, then is the tendency of the universe to flow from less likely (orderly) to more likely (disorderly) macrostates." This means that the pure, undetected

Information (Noise) is (as a whole/unity) they most orderly state; meaning is actually a result of entropy as "bits and pieces" are distilled/detected from the whole into particularities that are distinct and disparate. What it means to us and what it means to the universe are so different. Meaning/intelligence/knowledge are actually forms of entropy as the WHOLE is chipped away.

Quantum activity is possible because the O-field is seamless. "What if the O-field is a wave traveling through the universe at infinite speed making it present everywhere at once?"--William Sperier. My reply: That is one way to say it. But the word "travel" is tricky. Travel is relative to distance experienced in four dimensions. At the 0=1 level, there actually is no travel taking place because there is no separation. So we can experience and detect things as a wave in four-dimensional space that is "traveling" at an infinite speed, therefore not really traveling. It's in both places (technically ALL places) at once. Only observation and measurement give us the illusion or experience of something having traversed an actual distance. See...simple!

According to Schrodinger, "an organism sucks orderliness from its surroundings." Yet that indicates that entropy is more akin to meaning (Information being the purest form of zero entropy) and all forms of intelligence being dissipation from pure Information...the ALL.

"The gene is not an information-carrying macromolecule. The gene is the information." How is this entropy of Information reflected in the appearance/propagation of entropy on the information as it appears in the form of a gene?

"All these heroes of science were talking about and around randomness." It is random prior to detection and meaning.

"We would like to say that some numbers are more random than others—they are less patterned, less orderly." No! Numbers are arbitrary human constructs; therefore, they can NEVER be random. Information (Noise), however, can be.

Information in the universe includes not only the words spoken, but all thoughts thought.

"Chaos" is merely information prior to meaning.

Humans are not the sole detectors or "noise" and decipherers of "meaning." All conscious beings do this to varying extents. Even electrons themselves are detectors of meaning at a very basic level. Hence VERB.

"For in the end the disorder will become nearly insurmountable."—Alexander Pope. And yet THAT is what precedes meaning.

"Information is divorced from meaning." No. It simply precedes it as "potential" meaning.

Information precedes energy. Energy is just an expression of Information. The earliest and simplest expression of information taking form is energy. The changes of energy quantities expressed by a particle is also a result of Information.

Energy and mass, and the mathematics necessary to describe them, are all simply Information distilled and expressed in measurable quantities, phenomena, and formulas. Precedent to energy is pure Information, which is not measurable, but has a reality, and is itself the basis of all reality subsequent to non-distilled Information.

When any particle makes an energy leap from one quantity to another (as in the case of electrons), such a leap is result of Information. The particle distills or detects "instructions" from pure information and reacts in accordance to universal laws, governed ultimately by Information—the mind of God.

No, there is no god or God. There is only Information. From Information come all things known and unknown.

The initial state of the universe was pure Information. No meaning had been detected because there were no detectors. There was only pure Information in a state of zero entropy. With the Big Bang, some Information became energy, and later mass. Only then de we see the emergence of entropy in the universe. At the O-field level, entropy remains zero. Only at dimensions (measurements, sizes) larger than the O-field do we see interaction and detection, and therefore, increased entropy. Reality (existence, the universe itself), is merely the result of detection and distillation of meaning from pure Information. Distillation of meaning from Information, the moment of detection by a detector, is what happens in the collapse of the wave function of electrons. Prior to detection, the electron was more in a state of Information (potentiality) than in a state of meaning (actuality) for its detector, and therefore, for itself. Information and the O-field are indistinguishable from one another except in semantics and labeling. The Prime Will is also indistinguishable from Information and the O-field. They are merely three ways of describing aspects of the same phenomenon.

   Though the idea of zero entropy is plausible (if not definite), there is most likely no such thing as infinite entropy. Entropy can increase indefinitely, because, at the O-field level, Information is still in a state of zero entropy, regardless of what we experience in our four dimensions. Entropy decreases as we go "backward" in time and size. At the Big Bang, entropy was zero, and dimension and time were also zero. But Information was pure and undetected. No meaning had entered the universe yet. Only with the Big Bang do meaning and increased entropy emerge. Yet at the O-field level, the initial state of zero entropy remains intact. In other words, both are happening in the universe at the same time. Information remains pure and infinite at the O-field level, and entropy remains zero, while detectors in the universe (quarks, electrons, humans, black holes, galaxies) distill meaning and create reality from Information as a result of increased entropy at scales greater than the O-field. The prime mover in all this is the Prime Will, which is not an entity or a personality (like our concept of God), but merely an anthropomorphic term I have applied to the O-field/Information as an analogy for what, exactly, it is that is the driving force of change in the universe. The Prime Will moves all things forward to a higher state of entropy, engendering in its wake detectors like you and I. It is in this way that the old argument that the universe created intelligent life in order to have an observer and to reflect upon itself may very well have some truth to it.

So, what is Information? Not the classic dictionary definition. What the hell is it? Defined in terms of reverse engineering, "pure Information" is the teeming, vibrant force created when all of existence (mass and energy, gravity), is compressed into a singularity of $0=1$ dimension and $0=1$ duration. It is what it is. "I am that I am." At the Big Bang and soon afterward (forward engineering), we have all the forms that Information takes: energy, mass, gravity, temperature, work, galaxies, stars, planets, consciousness, and life. I am trying desperately to avoid metaphysics. I knows this encroaches dangerously near that realm.

Theorists are proposing and researchers are discovering the events a trillionth of a trillionth of a trillionth of a second AFTER the Bang. I am attempting to describe the Bang itself. It was when all of the known (and yet unknown) forces of the universe collapsed under gravity to a singularity of pure energy, and "collapsed" one step further to pure Information, before starting the process over again. We (may) inhabit one of an infinity of pulsating universes. All that remains constant is Information.

So, what again is Information?

That which IS when ALL that is is compressed to "is and is not"—$0=1$—THAT is Information.

"So, Information and the Prime Will are...?"

One and the same.

How Does It All End?

Astrophysicists have debated for decades: How will the universe end? Will it expand until there is nothing but cold, dead emptiness? Or will it conctract in what they call a Big Crunch?

I myself have been on the fence for over 20 years. But if $0=1$ is true as an aspect of reality, then I feel I prepared to get off the fence and make a claim:

The universe will end in a Big Crunch "followed" by another Big Bang.

I place "follow" in quotation marks because the time/duration "between" the Crunch and the Bang is equal to 0. The Crunch and the Bang are opposite sides of the ultimate 0=1 "coin".

But how, many will ask, if the universe is expanding at an exponentially increasing rate, is this possible?

Because, even as it expands, so too do black holes continue to grow by devouring stars within their galaxies. And then galaxies collide and the black holes at the center of each collide and devour one another, creating even larger black holes. With the growth of every black hole, its gravity increases. Since the whole of existence is connected via the 0=1 field, they are never separated. For a time, the power of the dark energy between galaxies is sufficient to keep them flying away from each other. But slowly, as galaxies (and the black holes at their centers), continue to grow, eventually their gravity becomes strong enough to overcome the energy separating them from distant galaxies. Their gravity becomes so strong, as they reach astronomical masses, that galaxies that were previously flying apart slow their escape, stop, and begin to converge toward another collision. Enough super black holes will collide and grow massive enough that their gravity will overcome even the most distant galaxies. In a frame of time that is almost incalculable, a "final" super black hole will grow massive enough to trap even the most distant and "final" rogue galaxy out on the fringes of the universe. Because, however far apart the black hole and the lone galaxy may be, the 0=1 field still keeps them connected. There is never not a pull of gravity between the two. Finally, the last vestige of repulsive dark energy will succumb to the overwhelming gravity of the final super black hole, which will stop the escape of the final lone galaxy and draw it to itself. With the devouring of the final lone galaxy, the final super massive black hole will contain all the mass of the universe. It will collapse in upon itself under its own gravity. Its event horizon, for a moment the largest thing in the universe, will shrink to a size smaller than a Higgs boson, all the way to nothingness, but with infinite mass. Space will cease to exist. Time will stop. Infinite mass will convert to infinite energy in a point/moment comprising zero duration and zero dimension. And then...Crunch/Bang! Simultaneously! 0=1.

Other *Other Implications*

What follows is a catch all for ideas that I jotted down as they occurred and I was unable, or too lazy, to find a fitting spot for them. So, here goes.

Spontaneous order in the universe should come to us as no surprise. If the Will to Equilibrium is correct, and the First Will is equivalent to the 0=1 point/moment of the Big Bang, then at that point everything in the universe was one. It all fits together spontaneously because at its origin it was already together. And in regards to the 0=1 field, it remains an inseparable unity. Even at the smallest levels, the puzzle pieces exist for one another and make a perfect fit.

The idea of the Will to Equilibrium, symmetry versus entropy, has some interesting implications for sexual reproduction and the act of sex itself. Somewhere between one and one and a half billion years ago, a single-cell life form experienced a genetic mutation that caused it to divide and split the information needed for reproduction into two separate entities, each one carrying half of the information. In order to reproduce, these tiny critters needed to "hook up" in order to share the information each possessed. For more on this aspect of sex, see Seife's *Decoding the Universe*. The initial division, the split, is an expression of entropy, an attempt to destroy. But Nature found a way to carry on. Sexual reproduction is an act of unity, of symmetry, two halves coming together to continue life. The division and the union, together, express the will to equilibrium. Fast forward a billion and a half years and consider human sexual reproduction. Many species seem to enjoy the sex act, but none more so than we humans. The joy and pleasure that comes from coupling is one of the greatest and most sought after human experiences. And the ecstasy and near loss of consciousness at the moment of orgasm, the "little death," as two become one and meld into a brief moment of oblivion, is a 0=1 moment, expressed through the climax of human intimacy. In the course of a human life, his or her electrons are never happier than at the moment of orgasm.

The reason all life forms on Earth share the same DNA and indicate a common ancestor is because our common ancestor was not a living organism that we all evolved from. Our common ancestor is DNA itself.

Space and time are NOT fundamental aspects of reality. Only Information, from which both space and time (space-time) are derived, is fundamental.

Physicists say it would take a particle accelerator the size of the universe and an energy amount that is infinite to prove the largest equations explaining the smallest phenomena in Nature. What, then, is it that we are living in? What do we observe when we look in any direction in the universe, and when we look inside ourselves? Nature herself is our particle accelerator and her energy is infinite.

Why is so much of life divided into two halves— what we call bilateralism? Why does our brain have left and right hemispheres? At some point, the single cell life form from which bilateral creatures evolved, had a genetic mutation that prevented it from dividing completely, leaving two halves with identical information still connected to one another. That mutation was replicated in a society that survived and carried on till today. Thus, even failed cell division itself, resulting in bilateralism, is an expression of 0=1, of symmetry versus entropy, of the pull to separate and the pull to remain one.

Gravity is constant, omnipresent, and pervasive. Electromagnetism is local and dependent on proximity and relative motion. Electromagnetism is "congealed" gravity. Experienced locally as electricity when motion is at play. Experienced locally as magnetism when stationary. Magnetism is not limited to so-called "magnets" But is present in virtually every element/particle to varying degrees and is synonymous with gravity, differentiated merely by degree. We know that magnetism and electricity are the same thing, thus the encompassing term electromagnetism. But since magnetism and gravity are also the same, therefore all three phenomena are essentially indistinguishable except by degree of influence on varying bodies as well as the structure and fabric of reality. Gravity is weakest because it exists as the 0=1 level. Electromagnetism does not appear until we reach the plus side of 0. Magnetism requires bodies however small, and electricity requires motion however slight. At the 0=1 level there are neither bodies nor motion, only the field that is omnipresent from white bodies and their motion are contrived.

0=1 is why acceleration and deceleration both encounter or require friction. Why we can accelerate. Why we meet resistance when we do. Why we can decelerate. Why we meet resistance when we do. Acceleration versus friction. Deceleration versus inertia. Symmetry versus entropy. 0=1.

Antimatter exists because of the Will to Equilibrium, which results from 0=1. Partner particles with a negative, attractive charge, and therefore the weaker of the two, are self-apparent in a reality originating in a 0=1 point/moment.

If 0=1, then the current dualities that still perplex investigators should be explained. The particle-wave duality; the location-speed duality; the mass-energy duality; and the magnetism-electricity duality–these are all examples of events that require an "in between" explanation, the point/moment when they are both and neither simultaneously, in a location and duration that are equal to 0, nonexistence. And yet, they contain the mysteries to all existence. This is the "trick" of reality that physicists and mathematicians have yet to discover. This is 0=1.

Changing the focus back to the everyday world with which we are familiar, it is interesting how the behavior at every level of consciousness is also reflected in the social interactions of all living creatures, including humans. Reactivity at the macro level? "I like this. I will keep doing it." "I don't like this I will stop doing it." At the quantum level, "I like this, I will keep spinning at this rate." "I don't like this, I shall speed up." Bonding at the macro level? "Hey I like you. Let's hang out. Want some food?" "Don't mind if I do." At the quantum level? "Hey, I like you. Let's hang out. Want an electron?" "Don't mind if I do!" And so, on up the chain of complexity and the spectrum of consciousness.

Considering free will in regards to the ideas I have presented here, it both exists and does not exist. Intention, as mapped in brain scanning, is often precognitive. The neurons "know" the next move before the subject is aware. This is because the initial activity is at the quantum level, in the motion of the electrons themselves, and occurs at or near the speed of light. Only after the quantum activity effects the chemical and neuronal activity do we become aware or conscious of our own thoughts and intentions can we speak of "free will" being in play, thus making "free will" an illusion on anything larger than the quantum level.

When using a term like "will" in reference to the universe–the Will to Equilibrium–I am not saying the universe itself is conscious. But I'm also not saying it's not. I think what we must do here is reexamine exactly what "will" is. In sentient beings, especially highly evolved, complex life forms, will is generally defined as the sum of our desires, as determination, as a particular stance is relation to an opposing force or entity. Will assumes choice in a given situation. But what is choice? And how much influence do we have over the choices we make? As said earlier, fMRI scans have detected electron activity in regions of a subject's brain indicating intent toward an action before the subject was consciously aware of their own thought. The intent was an electrochemical event precedent to actual awareness of desire or intent. Thus it, the intent, must be a result of quantum events hardwired by billions of years of evolution. Such activity seems to relieve us of what we consider to be our "will" if they are occurring without or prior to our conscious awareness. Does this mean we do not have will? Or does it mean we must expand the definition of "will" to include events that precede awareness. And if that is the case, can we not expand the definition of will to include entities that are non-living or non-sentient? Perhaps there is such a thing as pre-sentient, or proto-sentient. If such a thing were possible, then cannot the universe as a whole express or exhibit "will" while not being actually conscious or sentient? Clearly I believe it is not only possible, but inevitable. So when I discuss the "Will to Equilibrium" I am not saying the universe is making choices. I am saying it exhibits behaviors that are indicative of will, and with the expansion of the definition of "will" to include pre-sentient events, then the universe most definitely exhibits will.

Concerning art, the need to compose order on chaos is the will to equilibrium on a macro level. The angst and existential dread felt by most sensitive people, artists chief among them, is the push of entropy toward destruction from which, taking a cue from symmetry, they pull from their beautiful minds entire new worlds.

Another big question remains. Is there a -1 on the other side of 0=1, comprising a whole other universe in which what is -1 for us is +1 for them? It's possible. Maybe probable. But there is simply no way to confirm this. So we abandon the idea for now because it does not involve our universe, the reality we experience. It's none of our concern.

Also, though there is an absolute temperature on the cold end of the spectrum (0 degrees Kelvin), the point at which no more "work" is possible, therefore zero generation of heat, there is no such limit on the upper end. Therefore, the point at which pure energy reaches infinity, so too does the degree of temperature. They are one and the same. Heat, at this point, is also infinite. Existing at a point of 0=1 dimension and 0=1 duration, a point of non-existence, essentially, with infinite heat also suggests its instability and the resulting Big Bang. Infinite heat and infinite energy in a point/moment of 0=1 dimension and 0=1 duration. Existence at non-existence simultaneously. Infinity and absolute nothingness.

How we store memories: Sensory impressions, at the time of an event, are converted into chemical and molecular ingredients (namely dopamine, oxytocin, serotonin, and endorphins) that are stored and "catalogued" by electronic activity. These catalogues are capable of reproducing the impression in our "mind" when a new stimulus triggers a "memory" event. At the quantum level, electrons, moving near the speed of light, initiate a chain of molecular and chemical reactions that recreate the sensations imprinted and catalogued in our brains at the time of an event. They "play back the movie" of the experience. This is not to say there is a Cartesian theatre in our brains. Far more complex than that. When we experience a memory, our brain is essentially reliving the event. We are "seeing" and "hearing" again almost as vividly as we did in the initial event. The electronic, atomic, molecular, and chemical "record" of each impression and stimulus we experience is not so much stored like a book or video tape as it is recorded like a recipe that can be whipped together in fractions of a second and presented to our consciousness as a memory or revery. Like any recipe, however, the "preparation" "mixing" "baking" and "presentation" of the dish (memory) is never exact or precisely reproduced. Our brains can get very close. But 100 percent faithful reproduction is extremely unlikely. There are always alterations in the ingredients, leading to anything from slight misrepresentation of the impression to complete alterations and even fabrication of false memories. We can create new memories using the ingredients and processes that recreate actual memories of events and sensations we actually experienced. Memories of memories. Or memories of reveries. Or memories of daydreams. And they can all blend and cross and mingle and create a whole world of both novel and actual impressions. Sight and sound are almost exclusively the form in which memories occur. Rarely do we smell or feel or taste a memory. These involve neural pathways that are vastly longer and require many more steps to both create the initial impression and recreate the memory. The best we get is the emotional sense we had from the experience, as opposed to the actual sense impression. Although such things as tactile, olfactory, and gustatory memories can occur. Rare enough are they, and often unpleasant enough, as to render the memory as traumatic as the initial experience. Impressions of sight and sound are created and stored much faster and are likewise easier and quicker to reproduce. Our electrons, given their assignment by some new stimulus (which can itself be a memory) go to work at near the speed of light and "assemble" the team which assembles the ingredients which become the new memory in nanoseconds. The delay between the "assembly and baking" and our conscious awareness and experience of the memory follow mere microseconds behind. While the vast majority of experiences, due primarily to their humdrum nature (no one remembers every time they inhale and exhale) are never recalled, the ingredients for recall (electronic, atomic, molecular, and chemical) are all there nonetheless. For this reason we do, at times, recall the most random and seemingly insignificant events from our past, even from our early childhood. The brain, not just the human one, is almost indescribable in its awesomeness and power.

The concept of the O-field reconciles two previously at-odds theories: field theory and quantum mechanics. Which is the true theory? Is reality constructed of discrete quanta, or is it constructed of a seamless field? O-theory and the concept of the O-field satisfies both. A field comprised of a point/moment where 0=1 maintains BOTH discrete quanta AND a seamless, uninterrupted field. If there is a point/moment where 0=1 (as O-theory contends), then a field that is both seamless and discrete is possible, even probable.

How does O-Theory account for the four forces?

It is simpler than I thought.

They are the same phenomenon. The same force.

The only difference is scale. The smaller the scale, the stronger the force. The strong nuclear force (the strongest force) operates at the proton scale. Gravity (the weakest force) operates at the massive scale.

When we reach the 0=1 scale, all kinds of weird things occur. Time and dimension (space-time itself) cease to exist. Infinity and 0 are coexistent.

From the 0=1 scale to the scale of galaxies devouring one another, the forces are one and the same. There are not four forces to be reconciled. There is only one. It is the O-field itself. It is the universe. It is existence. It is reality.

"Conclusions"

This little thought experiment I have offered here can seemingly explain many, many more phenomena, if it is true. If. All you can do is try one for yourself. Think of a phenomenon at any level of the cosmos, from the biggest to the smallest, and see if the equation 0=1 in anyway could possibly be used to describe it. Ask yourself, in a world originating from, composed of, and exhibiting at every point/event in space and time the equation 0=1, could this equation explain your chosen phenomenon? This idea is not for me. I published it in a blog free of charge for all the world, a gift to humanity. Only later did I publish it in book form. "The worker is worth his wages." No? Assuming my idea may describe Nature and reality, make it your own, and see what possibilities you can imagine.

Before discarding my idea outright, consider that energy and mass were once believed to be separate entities. The idea that they are the same was absurd. Consider that it took sixty years for the Higgs boson to be observed. The theory was there and Professor Higgs and many of his colleagues "knew" it to exist. But the technology to prove it was sixty years in the future from when Higgs proposed it. The technology to prove 0=1, I suspect, may never be available. Computer models may one day "prove" it. But the size of equipment and amount of energy would be, well, the universe itself and all the energy contained therein.

I wrote this paper initially with only ten or twelve "implications." But 0=1 kept explaining more and more phenomena. I finally stopped listing additional implications in order to complete this paper and publish it. Even after publication, it has grown by over 5,000 words, over a dozen more implications. I could have kept going, indefinitely. If correct, this could literally be a Theory of Everything. If wrong, which is very likely, then consider it a cosmic joke, the brain child of a fellow of infinite jest.

VI. CLOSING THOUGHTS

I can't lie. I'm pretty damn excited about this little thought experiment and its potential impact on the world. It has met with confirmation after confirmation from my reading and in conversations with trusted friends. In my note writing and musings, every time I thought I may have coined a term or had a new concept, I would come across it in further reading and see it's already a concept or occurrence. That didn't bum me out. It got me even more excited because I knew I was on the right track. For example, I would speculate, "If such and such is true then the element that must be involved is…oh, nitrogen?" And I look it up and damned if nitrogen isn't superabundant in the environment I'm dealing with as well as the key ingredient in nucleotides. I guessed first and wrote a note about it, only to find it true down the road. Much of what I've written I can't call it "thought." More than half of this document came from notes written after waking from a dream in which these "illuminations" bombarded my mind. I would jot them down, thinking they were crazy, then I would check the science behind it and, lo and behold, it freaking worked! This has been my experience on this entire journey. I'm not a scientist, I just like to read and think. A lot.

The following was the original working title for this thought experiment:

AN UNSCIENTIFIC NARRATIVE SPECULATING ON THE ORIGIN OF CONSCIOUSNESS ON EARTH: Being a conclusion synthesizing cursory readings in several fields of study, including, but not limited to, anthropology, bacteriology, biology, chemistry, climatology, cosmology, crystallography, cytology, electrostatics, ethics, ethology, game theory, genetics, geography, geology, glaciology, history, information theory, meteorology, neurology, oceanography, paleontology, philosophy, primatology, psychology, physics, physiology, quantum electrodynamics, quantum mechanics, relativity, religion, sexology, sociology, statistics, thermodynamics, virology, and volcanology.

Those may not be all of the fields of study that contributed to the document I have created. They are simply the ones I could recall off the top of my head. My point in sharing this working title is merely to demonstrate that, right or wrong, my little thought experiment is the culmination of many years and an almost inhuman amount of reading. It was not an accident nor a fluke. It is the convergence of ideas contained in all of the fields of study listed above. It is my 0=1. Everything converging to one thing amounting to *no* thing explaining *every* thing.

Even if the ideas presented here prove to be wrong, I hope they are compelling enough to spark debate. I want to encourage thought, reflection, discussion, and progress. Before rejecting my ideas outright, at least consider them. Before laughing off and discarding 0=1, ask yourself, "Is it really impossible? Or is it just mathematically impossible?" The two may not be the same.

Today I was presented a challenge to my concept of 0=1. In a vodka-fueled social media post last night, I wrote: "I keep saying it: dark matter and dark energy are the same thing at the 0=1 level, the residual energy/matter from the Big Bang, at which energy was infinite, while dimension and duration were equal to 0=1.

Someday y'all will recognize." A friend commented that I had made an error in eliminating c from the equation. For a brief moment, I thought, "Holy crap. He is right. What about c?" In less than two minutes, however, I had figured out the problem and had a very definite "Duh!" moment.

I replied:

If I am correct, then c is not required at the 0=1 point/moment. Einstein's theory is applicable for measurable quantities of mass/energy and their equivalence, in space, with c as part of the equation. 0=1 is THE conversion point/moment at which dimension (think "size") and time (duration) are both equal to 0. Time and space essentially do not exist at that point. So the speed of light is irrelevant. Since, I believe, the amount of energy at the Big Bang was infinite, that energy remains infinite, even though much of it has converted to matter. So what we have is a universe that is roughly 4.9% normal matter, 26.8% dark matter, and 68.3% dark energy. [Physicists came up with those numbers by tinkering with computer models until they got a computer-generated universe that looks like our telescope photos. So it's all guesswork.] That last number is misleading because the energy amount is actually infinite, so it can't have a terminal percentage. It's more accurate just to say "all the rest of the universe" is dark energy. I'm arguing that all of that 95% dark matter and dark energy (combined) is, at the 0=1 point moment (which is the "field" of the universe itself), really the same thing, with approximately 25% exhibiting the properties of matter (like gravity) and the other 70% exhibiting the properties of energy (like accelerating expansion of the universe).

I know I am fighting an uphill battle. It really would not take much to convince me I'm wrong. And pointing out potential errors like my friend did is the way to do it. Eventually I'm not going to have any response other than, "You know what? You're right!" In the meantime, challenges make me go, "Let me think about out it for a minute." This paper was originally fourteen pages long. As of this writing it has topped 88 pages. Those additional 74 pages (and counting) are the result of challenges and questions. People have requested clarification on several points, and much of what has been added was simply elaboration. More importantly, several readers have challenged a point and forced me to ask, "If I am right, how do I explain such and such?" And each time, thus far, the result has been more words. I ponder the problem, usually for no more than a couple of minutes, and the answer comes to me. When the day comes that I cannot answer or explain a phenomenon in relation to 0=1, I will gladly concede my error and discard the hypothesis.

When confronted with the seeming impossibility of what I have proposed here, I usually reply, "Correct!" I'm not denying its "impossibility." But relativity was also once "impossible." Quantum entanglement was one "impossible." Gravity itself was once "impossible." You should read, if you haven't already, some of the accusations and insults against Isaac Newton for DARING to suggest some "force" between entities in the universe was pulling them together toward a "central" point. Crazy! Impossible! But he was right. My idea is also "crazy" and "impossible." But I'm convinced I am right, until convinced I'm wrong.

For a century we have been locked, trapped, and imprisoned by this notion that mathematics is the only language in which Nature speaks. We fell into this fallacy when Newton discovered that mathematics reflects Nature. When Newton proposed his equations and they proved, time and again, to be correct, we said to ourselves, "Eureka! The language of Nature is MATH!" Nature, like much of mathematics, is consistent. But mathematics is not complete. Nature retains a few tricks up her sleeve that do not translate into the language of mathematics. Mathematics is a human construct. I don't know how many times I have to point this out. Yes, much of nature and mathematics reflect one another. But remaining locked in a mindset that Nature cannot be described except through mathematics is the reason we have not moved forward in our discovery of her deepest mysteries in over 100 years. My concept of $0=1$ flies in the face of conventional mathematics. And it could very well be as incorrect and ludicrous as it has been labeled.

But until we shake free of the notion that mathematics and mathematics alone hold the keys to the secrets of Nature, she will remain elusive. We must ask bold questions again. We must propose ludicrous theories. And we must have the bravery to pause and ask ourselves, "What if?"

What do I hope to accomplish by sharing this document with the world. I hope, more than anything, that my little thought experiment, even if it remains unproven, or is disproven, can be somehow helpful to humanity. Cliché, perhaps, but sincere. Hear me out. If my idea is true, maybe it will aid in the arrival of a quantum computer. As long as mathematicians use binary computers, 0's and 1's, products of a post-$0=1$ reality, they will not be possible. Until the equation $0=1$ is part of their equations, it will not happen. Even qubits, as clever of an idea as they may be, are insufficient. Although the closest anyone has come to describing reality at its most fundamental level, qubits are still the product of humans attempting through mathematics and abstractions to do what, so far, only Nature can do–make $0=1$. In regards to information technology, access to and utilization of the $0=1$ "corridor" aspect of the O-field could allow for the exchange of literally unlimited amounts of information among potentially countless contact points at an infinite speed in a duration of time equal to $0=1$. And why stop with technology? Perhaps there are quantum solutions to mental illness and intellectual disabilities, ways to help individuals stalled on a lower rung of the spectrum of consciousness to rise a little higher. Maybe there are quantum cures for physical illnesses and cancer. Perhaps there are quantum methods to increase creativity and accelerate learning. I can imagine quantum methods devised to increase morality, decrease aggression, increase altruism, and help with cravings and addictions to substances like tobacco and alcohol. I do not mean pills and chemicals with horrible side effects. I am talking about pinpointing the electrons themselves. With 0=1 technology, we will eventually be able to pinpoint the very electrons at the base of such chemical activity directing ailments, mental disorders, and possibly even physical disabilities. Technology stemming from O-theory is the key to the next generation of brain mapping and exploration. Ultimately, increased resilience for our species as a whole could be a possible future application, utter immunity, even immortality—with ethical issues always looming. All of these human realities are results of electronic behavior, electrons seeking equilibrium. Happy electrons. Surely there will one day be technology to assist the electrons in their pursuit of happiness. I really do not know what is possible. I only know I dream big.

One more question: Is human intellect the apex of evolution in the universe? Is Global Awareness the highest rung, the culminating point of the spectrum of consciousness? Is there an observer-quantifier-qualifier who observes-quantifies-qualifies more precisely and deeply than we do? Our tools of measurement have grown in complexity and reach, but has our ability to process information actually grown, and can it develop any further? Data has increased a trillion-fold (a gross understatement), but have we grown any more adept and potent in our ability to perceive and "know"? If we have—with quantum mechanics and theoretical mathematics delivering to us "knowledges" that press the bounds of reality itself—how much further can we go? If we have not, is it because there is no further to go? Are we it? It would seem we can expect only more discovery to be made, not evolution in the process and facility, or capability, of knowing. If beings exist who have achieved a mode of perceiving and knowing that is significantly advanced compared to ours, why have we not heard from them, except for the sheer vastness of space? How many "leaps" in evolution to go till we achieve intergalactic, or even trans-universal, communication?

I cannot help but feel an overwhelming awareness of history and my possible place in it. I could be no more than a nameless, faceless jester lost to time and memory. Or I could find my name placed alongside those we consider giants of human intellect and discovery. I can hear the laughter now, and I'm not averse to good-natured ribbing or vicious and deliberate ridicule. I'll take whichever the paper deserves when its impact, or lack thereof, is finally known. But I'm reminded of the voices of the scientific past who once were mere dreamers, crackpots, lunatics, cranks, quacks, and madmen whose intrepidness and foresight eventually placed them in the annals of history and made their names those which later generations of explorers would whisper in reverence. For each of them, their ideas began as pipe dreams and, for their less-visionary contemporaries, absurdities and impossibilities. And yet, today, what were their fantasies are our modern-day facts.

Just as Julius von Mayer and Hermann von Helmholtz, to name only two, had their respective papers rejected by the leading journal of their time, I have shared my paper with the likes of consciousness explorer and journal publisher Huping Hu, MIT physicist Jeremy England, philosopher of mind Daniel Dennett, and historian Peter Watson, all to no avail. Only Dr. Hu graced me with a rejection letter. The others? Silence. I even took a shot and sent the link to former President Barack Obama and tweeted it to astrophysicist Neil deGrasse Tyson. A man can dream. Whether the ideas I've proposed here will become scientific law or be shuttled into the dustbin of harebrained ideas is yet to be seen. I may not live to see the outcome. In the meantime, I will keep dreaming.

The intellectual journey I have taken, of which this paper is the result, at times brought me to the edge of sanity. Meddling with the mysteries of the universe and the inevitable impossibilities encountered in such endeavor can tax the mind of any human. Perhaps the faith I have come to place in the equation, $0=1$, was a result of me folding under mental pressure, a coping mechanism to deal with the mental strain I had put myself under. Maybe I cracked. But I think not. The equation, if true, explains too many phenomena–everything, actually–for it to be the fever dream of a brain stretched to its limits. I accept is as a fundamental aspect of reality, at least until something more convincing comes along.

There is something to be said for the certainty intellectual explorers have when they "know" they have discovered something significant. Planck said, "I have had a conception today as great as the kind of thought Newton had" (Brennan 94). Regarding the structure of the atom, James Chadwick said of Rutherford's revelation: "We realized this was obviously the truth, this was it" (Rhodes 50).

At the end of 2014, Joseph Silk and George Ellis wrote in Nature, "some scientists appear to have explicitly set aside 'the need for experimental confirmation of our most ambitious theories, so long as those theories are sufficiently elegant and explanatory.' They further complain they were we are at the end of an era, 'breaking with center is a philosophical tradition,'" the defining scientific knowledge as empirical. O-theory may not be empirical now, but with time it will become so.

I wish I was actually working on this stuff and trying to figure it out instead of my brain BLINDSIDING me with thoughts and images and ideas I have no say in other than to write them down as quickly as they come.

When I was 17, and a quite zealous Christian, I believed I heard the Lord whisper to me, "Ask anything you want and I will give it to you." Recalling the story of Solomon, who I admired above all other characters in the Bible, except for Jesus, I said, "I want wisdom, Lord." But I went a step further and said, aloud, "I want to know everything you know." And the Lord said, "Study." And thus I began a fervent pursuit of knowledge in college and in personal research. Ironically it was that knowledge which I asked for of God that led me, rather quickly, to realize he did not exist.

Having rejected the God hypothesis long ago, I realize now I can never know the mind of a being that does not exist. Such an idea is just as frivolous as believing I can know everything there is to know. I have struck a happy medium and sought to know everything I possibly can and continue to learn every day of my life.

A curious observation: Interestingly, we can see the whole of the spectrum of consciousness exhibited in the development of a human from birth. Although already a very complex being, much of a newborn's behavior resembles Reactivity (in a complex, living being we call it "reflex") and Bonding, as in the emotional connections it makes with its mother, father, and other persons in its young life. An infant at birth has already reached the level of Instinct. From there a child develops through the other levels, sometimes stopping somewhere along the way, other times making it to the uppermost level. What lies beyond for humans, if anything, is yet to be seen.

So, what really happens when we die? Earlier in this paper I listed and defined the ten degrees of consciousness. For times' sake I will list them again: Reactivity, Bonding, Instinct, Essential Awareness, Cognition, Permanence, Intention, Foresight, Self-Awareness, and Global Awareness. Put briefly, when we die, our bodies go through the spectrum in reverse. Depending on what degree of consciousness we are on, the process may involve fewer than ten steps. Single-cell organisms clearly have fewer levels to recede through on their way to oblivion. For the purposes of this discussion, I will focus on what happens to *homo sapiens* at this moment we all one day face. If the death is sudden and catastrophic, like incineration, several, nearly all, levels are passed through simultaneously. It all happens at once. However, let's consider a slower death, one in which we pass through each level distinctly. The first thing we lose is global awareness, awareness of others around us. We a left momentarily with the awareness of only ourselves, our single voice trapped in our head, trapped in isolation and loneliness. Thus confirming Conrad's admonition: "We die as we dream. Alone." It is into this private dream world we slip as we lose awareness of our surroundings and know only ourselves. From there, self-awareness abandons us. For those who are hoping to be conscious of themselves and their personality in some form of an afterlife, where they will meet loved ones who have died before them, I have some disappointing news. Global and self-awareness do not survive death. They are the first to go. And unless, after the process is complete, our energy reconstitutes itself in another dimension and passes back through all of the levels, there is no afterlife in which we know and recognize ourselves and one another. But, to continue: once we lose self-awareness, the process accelerates. Foresight, Intention, and Permanence flee quickly, possibly all at once. We are left with a crude form of cognition of which, without the higher, more-complex levels, we are hardly aware. Essential awareness remains in that our senses are still receiving, and our brain still processing, external stimuli, though we make not hide nor hair of them. Near what most of us consider "the end," we lose instinct. To the last moment, we blink, we breathe, we swallow, we flinch when prodded. In a final grasp at survival, our gross motor skills and involuntary functions are still at work. Then, darkness. Essentially, we are dead. We can be buried and mourned. Our loved ones begin to speak of us in the past tense. And yet, there remain two degrees of consciousness that are still at play. Our remains travel in reverse through Bonding, as our tissues, membranes, molecules, and atoms begin to decay. We rot. We are absorbed into the ground or evaporate into the air. Some part of us is consumed and used as energy by plants and worms and fish and birds, depending on where and how we meet our demise. Finally, all that is left is our energy, the electrons dancing at the level of Reactivity. Where do they go? Does the decay continue through the quark level, on down to the maxiquantum level, until what once was us reaches the $0=1$ level? Are we then assimilated back into the very fabric of the universe? It may be. So, for those dreaming of an afterlife, perhaps there is a form of immortality. Though far from what we envision or what we were told in church or mosque, perhaps we do endure in some form. I would say it is not only possible, but probable—maybe even certain.

Moving backward through the many studies necessary to comprehend consciousness in general and human consciousness specifically, be begin in a vast field of all the disparate and possible areas of knowledge built upon the earlier and more established fields: all of the fields find themselves based on biology, then chemistry, then physics, and finally mathematics. It is just beyond this point that $0=1$ takes us.

On that note, I would be remiss to bring this document to a close without addressing faith, or more particularly, belief, one more time, be it faith in a deity, or faith in an impossible idea. Imagine a world in which we were truly free to believe or not believe and all get along because both sides had proof, and therefore assurance, of their belief? I may have just provided it. The order we see at all levels of reality, from the bonding of atoms to form molecules, through the formation of complex societies, could be the handiwork of some sentient, cosmic being. Considering spontaneous order and the mechanism of natural selection, such a being becomes extraneous. For those inclined to believe in God, my little thought experiment, especially the equation $0=1$, proves he exists. God *is* $0=1$, the Prime Will. If you are inclined not to believe in God, my little thought experiment proves him unnecessary and therefore he most likely does not exist. The choice is up to you.

Finally, as much as I detest pseudoscience, it may turn out that my little thought experiment is nothing more than that. It's definitely not science in the sense that I've provided no experimental evidence; nor is it philosophy. To be honest, I'm not sure what it is. All I know is I had hoped to offer an elegant idea. I also had hoped to write an elegant piece of prose worthy of the magnitude of the idea it contains. My fear is I have failed on both counts. Time will tell.

BIBLIOGRAPHY

Works not specifically cited in the text are included here for their overall influence and inspiration.

Adams, Douglas. *The Ultimate Hitchhiker's Guide to the Galaxy*. Del Ray, 2002.

Aflalo, Tyson, et al. "Decoding motor imagery for the parietal complex of a tetraplegic human." *Science*. 22 May 2015. 348:6237. 860-861 and 906.

Anderson, Phillip Warren. "More is Different: Broken Symmetry and the nature of the hierarchical structure of science." *Science*. 4 August 1972. 177: 4047. 393-396.

Blake, William. *The Complete Poetry & Prose of William Blake*. David V. Erdman, editor. Anchor, 1982.

Bohm, David and B.J. Haley. *The Undivided Universe: An Ontological Interpretation of Quantum Theory*. London: Routledge, 1993.

Bohr, Niels. *Atomic Physics and Human Intellect*. Dover: Dover Publications, 2010.

Brennan, Richard P. *Heisenberg Probably Slept Here: The Lives, Times, and Ideas of the Great Physicists of the Twentieth Century*. New York and Chichester: Wiley, 1997.

Buonomano, Dean. *Your Brain is a Time Machine: The Neuroscience and Physics of Time*. Norton, 2017.

Campbell, Joseph and Bill Moyers. *The Power of Myth*. Doubleday, 1988.

Chadarevian, Soraya de. *Designs for Life: Molecular Biology after World War II*. Cambridge: Cambridge UP, 2002.

Davies, P.C.W., and J. Brown, editors. *Superstrings: A Theory of Everything?* Cambridge: Cambridge UP, 1992.

Dennett, Daniel C. *Consciousness Explained*. Boston: Back Bay Books, 1992. Print.

Dennett, Daniel C. *Darwin's Dangerous Idea: Evolution and the Meanings of Life*. Simon and Schuster, 1996.

Dennett, Daniel C. *From Bacteria to Bach and Back: The Evolution of Minds*. Norton, 2017.

Fortey, Richard. *The Earth: An Intimate History*, London: HarperCollins, 2004.

Geertz, Clifford. *Local Knowledge*. New York: Basic Books, 1983.

Gleick, James. *The Information: A History, a Theory, a Flood.* New York: Vintage, 2011.

Goodall, Jane. *Through a Window: Thirty Yeas with the Chimpanzees of Gombe*. London: Weinfield & Nicolson, 1990.

Greene, Brian. *The Elegant Universe: Superstrings, Hidden Dimensions, and the Quest for the Ultimate Theory*. Norton, 2010.

Greene, Brian. *The Fabric of the Cosmos: Space, Time, and the Texture of Reality*. Vintage, 2005.

Greene, Brian. *The Hidden Reality: Parallel Universes and the Deep Laws of the Cosmos*. Vintage, 2011.

Gribbin, John. *Q is for Quantum: Particle physics from A to Z*. London: Phoenix, 1998.

Harris, Sam. *The Moral Landscape: How Science Can Determine Human Values*. Free Press, 2011.

Hawking, Stephen. *A Brief History of Time*. Bantam, 1988. .

Hawking, Stephen and Leonard Mlodinow. *The Grand Design*. Bantam, 2012.

Holt, Jim. *When Einstein Walked with Godel: Excursions to the Edge of Thought*. Farrar, Strauss, and Giroux, 2018

How the Universe Works. Discovery Channel and The Science Channel. 2010-2015.

Hunt, Morton. *The Story of Psychology, Updated & Revised Edition*. Anchor, 2007.

Hunt, Tam. "The Hippies Were Right: It's All about Vibrations, Man! A new theory of consciousness" Observations, *Scientific American*, 5 December 2018.

Isaacson, Walter. *Einstein: His Life and Universe.* Simon and Schuster, 2007.

Kaku, Michio. *Einstein's Cosmos: How Albert Einstein's Vision Transformed Our Understanding of Space and Time.* Norton, 2005.

Kauffman, Stuart. *The Origins of Order: Self-Organization and Selection in Evolution.* Oxford: Oxford UP, 1993.

Lane, Nick. *The Vital Question: Energy, Evolution, and the Origins of Complex Life.* New York: Norton, 2015.

Laughlin, Robert B. *A Different Universe: Reinventing Physics from the Bottom Down.* New York: Basic Books, 2005.

MacDougall, J.D. *A Short History of Planet Earth*, New York: Wiley, 1996.

McFadden, JohnJoe, and Jim Al-Khalili. *Life on the Edge: The Coming of Age of Quantum Biology.* New York: Broadway Books, 2014.

Monod, Jacques. *Chance and Necessity: An Essay on the Natural Philosophy of Modern Biology*, New York: Knopf, 1970.

Nagel, Ernest. *The Structure of Science: Problems in the Logic of Scientific Exploration.* Indianapolis and Cambridge: Hackett, 1979.

Oppenheim, Paul and Hilary Putnam. "Unity of Science as a Working Hypothesis," in Herbert Feigl, Michael Scriven and Grover Maxwell, eds, *Concepts, Theories and the Mind-Body Problem*, Minneapolis: University of Minneapolis Press, 1958.

Penrose, Roger and Stuart Hameroff. "Consciousness in the Universe: Neuroscience, Quantum Space-Time Geometry and Orch OR Theory." Penrose, Roger, et. al., editors. *Quantum Physics of Consciousness*. Cosmology Science Publishers, 2011.

Prigogine, Ilya and Isabelle Stengers. *Order Out of Chaos: Man's New Dialogue with Nature*. London: Flamingo/Fontana, 1984.

Rhodes, Richard. *The Making of the Atomic Bomb*. New York and London: Simon & Schuster, 1988.

Russell, Bertrand. *A History of Western Philosophy*. Simon and Schuster, 1967.

Sagan, Carl. *Cosmos*. Random House, 1980.

Sagan, Carl, Ann Druyan, and Steven Soter, writers. *Cosmos: A Space Time Odyssey*. 20th Century FOX. 2014

Schrodinger, Erwin. *What is Life?* Canto Edition. Cambridge: Cambridge University Press, 1992.

Science. 22 May 2015. 348:6237. 860-861 and 906.

Seife, Charles. *Alpha and Omega: The Search for the Beginning and End of the Universe*. Penguin, 2004.

Seife, Charles. *Decoding the Universe: How the New Science of Information Is Explaining Everything in the Cosmos, from Our Brains to Black Holes*. New York: Penguin, 2006.

Shakespeare, William. *The Complete Works of William Shakespeare: All the Plays, All the Poems*. 2 vols. Doubleday, 1970.

Simpson, George Gaylord. *This View of Life: The World of an Evolutionist*. New York: Harcourt Brace and World, Inc., 1964.

Soni, Jimmy, and Rob Goodman. *A Mind at Play: How Claude Shannon Invented the Information Age*. New York: Simon and Schuster, 2017.

Taylor, John. *Hidden Unity in Nature's Laws*. Cambridge: Cambridge UP, 2001.

Through the Wormhole with Morgan Freeman. The Science Channel. 2010-2016.

Tyson, Neil DeGrasse and Donald Goldsmith. *Origins: Fourteen Billion Years of Cosmic Evolution*. Norton, 2005.

Watson, Peter. *Ideas: A History of Thought and Invention, from Fire to Freud.* Harper Perennial. 2006.

Watson, Peter. *The Age of Atheists: How We Have Sought to Live Since the Death of God.* Simon and Schuster. 2014.

Watson, Peter. *Convergence: The Deepest Idea in the Universe.* Simon and Schuster. 2016.

Watson, Peter. *The Modern Mind: An Intellectual History of the 20th Century.* Harper Perennial. 2002.

Weinberg, Steven. *The First Three Minutes: A Modern View of the Origin of the Universe,* New York: Basic Books, 1977.

Wikipedia contributors. More articles than you can shake a stick at. *Wikipedia, The Free Encyclopedia.* Wikipedia, The Free Encyclopedia, 15 Sep. through 15 Nov. 2016. Web.15 Sep. through 15 Nov. 2016.

Wolchover, Natalie. "First Support for a Physics Theory of Life." *Quanta Magazine Online.* 26 July 2017. Accessed 15 October 2018. https://www.quantamagazine.org/first-support-for-a-physics-theory-of-life-20170726/

Wolfram, Stephen. *A New Kind of Science.* Champaign, Illinois: Wolfram Media, Inc., 2002.

Zaugg, Julie. "The 'blob': Paris zoo unveils unusual organism which can heal itself and has 720 sexes."

CNN.com, CNN, 17 October 2019,
     https://www.cnn.com/2019/10/17/europe/france-
new-organism-zoo-intl-scli-scn-hnk/index.html.